Wings in the Light

Wings in the Light

WILD BUTTERFLIES IN NORTH AMERICA

David Lee Myers

Foreword by Robert Michael Pyle

Yale UNIVERSITY PRESS | NEW HAVEN AND LONDON

Frontispiece: Hoary (Zephyr) Anglewing, Challis
National Forest, Idaho

Yale University Press books may be purchased in
quantity for educational, business, or promotional
use. For information, please e-mail sales.press@yale
.edu (U.S. office) or sales@yaleup.co.uk (U.K. office).

Designed by Nancy Ovedovitz and set in Sabon and
Verlag types by BW&A Books, Inc. Printed in China.

ISBN 978-0-300-23613-2 (hardcover : alk. paper)

Library of Congress Control Number: 2018946730

A catalogue record for this book is available from
the British Library.

This paper meets the requirements of ANSI/NISO
Z39.48-1992 (Permanence of Paper).

10 9 8 7 6 5 4 3 2 1

To everyone who goes outside and looks around

To Alexandra, who goes outside with me

ABOUT THE XERCES SOCIETY FOR INVERTEBRATE CONSERVATION

The Xerces Society is a nonprofit organization that protects wildlife through the conservation of invertebrates and their habitat. Established in 1971, the society is a recognized leader in the protection of butterflies, bees, and other pollinators, and has become a trusted source for science-based information and advice. With the largest and most experienced pollinator conservation team in the world, we collaborate with people and institutions at all levels and our work to protect pollinators encompasses all landscapes. Our team draws together experts from the fields of habitat restoration, entomology, botany, farming, seed production, and conservation biology with a single focus—protecting the life that sustains us. To learn more about our work, please visit www.xerces.org.

Wings in the Light is a delightful book that introduces readers to the beauty of butterflies and the simple pleasures from time spent watching them. From that grows love, and from love a desire to care for and protect. The Xerces Society is proud to support publication of this book, and encourages anyone who picks it up to move beyond reading and into action!

Contents

Foreword

by Robert Michael Pyle

It is always more fun to have a co-conspirator, a comrade, a colleague, call it what you will, but someone who shares your esoteric interests and enthusiasms when it comes to a hobby.

Now, let me back up already. There is nothing inherently esoteric about butterflies. They are, after all, one of the most universally beloved entities on the planet. Almost no one dislikes them, quite of lot of people actively love them, and an interest in their well-being is commonly shared. And back up a little farther: butterflies are not really my "hobby." You could have called them that in fifth grade, when I first took my embryonic collection to school in a cardboard box. But from not so long after that, and for much of the sixty years since, butterflies have been a major part of my life's engagement with the world, and at times even my profession.

Even so: chasing butterflies and studying their distribution in my home county and state are what I do for fun these days. And there aren't that many people who share this pursuit in the detail with which I pursue it. So it was a keen pleasure to me when, twenty-five years ago or more, an acquaintance-becoming-friend in the next valley over from mine took up a butterfly net and a practiced camera, and joined the hunt in high gusto. That person was David Lee Myers, the author and photographer of the extraordinary book you hold in your hands.

To have an intellectually stimulating neighbor (eight miles away as the raven flies, twenty by country road), whom I also happened to personally like, take up the study of butterflies and begin to seek their fair portraiture through his professional lenses, was both a boon and a treat for me. Even more so when he took over as organizer and data keeper of the Western Wahkiakum County Fourth of July Butterfly Count, our local annual inventory in the maritime rain forest of southwestern-most Washington. Here in the Kingdom of Rain we entertain rather few species, but the well adapted ones can be numerous, and who can complain about a day out among scads of cherry, ebony, and mica-colored Clodius Parnassians, especially with someone else keeping track of the count?

As the years went by, and David's knowledge of the butterflies increased, he took himself farther afield to encounter species with which we of the northern forests are ungraced. And everywhere he went, he took his cameras, already distinguished for their treatment of forest, shore, and other Northwest landscapes. I know the challenges of butterfly photography, having pioneered the field with simple equipment in the sixties and seventies, and published the first field guides illustrated with photos of live butterflies. But I put the camera down with the digital revolution, and David continued, to become a true master of the tricky mix of electronic pixels with the pixilated scales of the bright wings of summer. From the start, he was a pro's pro.

Which brings us to this book, *Wings in the Light: Wild Butterflies in North America*. There are many butterfly books out there. Some are aimed at identifying every species you might encounter. Others deal with the anatomy, biochemistry, and molecular biology of their relationships and lifeways. Still other titles attempt to acquaint readers in detail with the life cycles and metamorphoses, ecology and conservation challenges, or evolutionary biology of butterflies. None of these are Myers's particular concern. If anything, I would say that he hopes to present butterflies to the reader and viewer as integral members of the same community of life that we all occupy; and as elements of diversity in the landscapes we all frequent, attending to which can make our outings and our very lives much the richer and more rewarding.

Now, the information presented here is reliable and notably broad in its application, so anyone who attends to the text as well as the wonderful images will come away with a basic education in butterfly natural history. David's familiar and confiding style only enhances the book and its lessons' accessibility. You'll find a lot of facts here to address questions you may have been wondering about, concerning how butterflies live their lives. As well, you'll find many suggestions of resources for enhancing and informing your activities with butterflies. David also shares his own secrets as to how you might become more adept in your own pursuit of butterfly photography. The text is charming, and substantial, far from mere captions for the stunning photographs. And stunning they are: I do not know a butterfly book with more beautiful images of the admittedly brilliant subjects.

I am now going to admit to one of my own biases and how it pertains to my delight in David Lee Myers's images. Butterflies do not simply emerge from their chrysalides to lead blameless lives of ethereal, ephemeral perfection. They live life! And they are subject to weather, predators, aging, and any number of other adverse factors, just as we are. Yet if you depend for your impression of these creatures upon most photographers or artists, you would think that all butterflies are mint-fresh, all of the time, cradle to grave, or chrysalis to dust. Most picture-takers, and hobbyist collectors for that matter, reject the torn wing, the rubbed colors, the lost tail, the old and worn veterans, as if they didn't even exist.

Canny museum curators know better: specimens admitted to their trays should illustrate all the vicissitudes that the organisms have weathered to get there; they have, after all, lived real lives! And those are interesting lives, full of adventure: encounters with birds, crab spiders, wind and bushes, time, and all the other aging and weathering factors the environment can bring. David Lee Myers is perhaps the only butterfly photographer I know who fully honors the life challenges of the insects he captures on "film." Swallowtails without tails, an expired pine white, a worn-out Callippe fritillary, an arctic fritillary missing most of its hindwings—these are the stuff of *True-Life Adventure* in the butterfly realm, and I greatly admire Mr. Myers for paying them their due attention.

And here's another personal prejudice. When I was a graduate student in New Haven, Connecticut, I loitered in a premise down the street from Mory's known as the Myron Gallery. Somehow the proprietor had come into a consignment of a discontinued batch of Leitz binoculars that he sold at such a discount that one pair became my butterfly- and bird-watching tool for the next forty years. But the other thing I most remember from that shop was the array of bird paintings by Robert Verity Clem. I couldn't afford any, but eventually I bought his big book on the shorebirds of North America. One thing I particularly loved about Clem's renderings was the way in which he sometimes depicted the subject bird as almost incidental in the landscape—a sandpiper among trees and sedges, a snipe snuggled deep in a canvas of reeds and rushes. I loved these paintings, causing one to look deep into the scene to discern the subject, almost as in nature. Once more I thought of this as I scanned David Myers's photographs, because he sometimes does the same: a painted lady buried deep in grasses and leaves, an admiral way up atop a marah vine, a zephyr anglewing tucked all but invisibly among dead branches. This is daring stuff, for any photographer, to hide the subject out of plain view.

Of course, there are also the many full, fresh portraits of beauties up close, basking, nectaring, even in flight. I don't know any photographers who capture butterflies better through their cameras. And other kinds of depictions as well: some show real wit, as with a northern blue peeking over a leaf, forewings akimbo; or a remarkable snout butterfly poking its rarely pictured proboscis (and its big palpi, or snout) through a forest of arm hairs for a sip of salty sweat, as photographed by David's brother Fred.

Behavioral tableaus, such as courting and mating and feeding, also include an excellent section on predation. The most shocking shot caught just the wing tips and antennae protruding from the maw of a lucky lizard. A number of deeply affecting and delightful portraits of people and butterflies together pop up throughout. My favorite of course is a lovely full-face smile of my late wife, Thea, graced with a checkerspot on her cheek. And I can't help but smile at one of an agile young lad following me at a distance, each with our net, both of us dwarfed by towering firs above the edge of the meadow. The various butterfly faces attest to the utter delight to be found out among the butterflies in their Elysian fields.

One of David's most original images is a haunting bilateral monarch-wing-and-eucalyptus construction that he calls "Mask of the Monarch." I've seen nothing like it, and it will stay with me always. And speaking of monarchs, he includes an excellent section on *Danaus plexippus* that draws partly from a memorable trip that we shared to their winter home in Michoacán. If you can't go see the monarchs in their overwintering spectacle yourself (and this will encourage you to do so), you can get an excellent sense of the experience from these glittering images.

But again, it would be wrong to suggest that this is just a magnificent picture book. The text, built from personal experience and solid research, is at once familiar, unassuming, confiding, witty, and humble, yet reassuringly authoritative—a difficult balance to achieve, but one that comes naturally to David Myers. Readers receive a pictorial introduction to all the families of butterflies they will encounter, and this is especially valuable for those who wish to get to know these insects better. The book finishes with a thoughtful take on my own major area of concern, butterfly conservation, set in a larger environmental context—concise, informed, and originally expressed. He manages to give us hard-headed hope, a tall order these days.

On the whole, I can't say enough about *Wings in the Light*. I couldn't be more delighted that my old friend and field compatriot, David Lee Myers, has realized such a brilliant, beautiful, and satisfying expression of his deep passion for our shared love, the wild butterflies.

Preface

Butterflies are a visual treat. And a bit of a tease: often too far away and too skittish to easily see. With patience or chasing, good binoculars, cameras, or nets we can see and study them more closely. Butterflies looming overhead, some eye to eye with the lens, others perched on people—all different ways of being in their world—butterflies in the wild, seen in natural light and shadow. Photographs taken in these situations began the work that became *Wings in the Light*. Readers attracted by these photos will also find introductory biological information here, in language that's both accessible and scientifically correct, and well illustrated.

Butterflies are sensitive to environmental disruption and climate change, so monitoring butterfly populations is an effective way of tracking the effects of those changes. This is an area where citizen-scientists contribute, and I hope more will join the effort. Anybody who has learned birding, botany, or other nature study has the required skills and attitude. They know the drill: Go outside and look around. See something, try to figure it out in the books. Not sure. With luck, see it again, this time a little more thoroughly. A few rounds of this and now one knows, and can move on to the next puzzle piece. Gradually the picture is assembled.

Unlike field guides, my book is an overview of beauty and fascination. Details are introduced, and then readers are directed to the books that do give the comprehensive lists and finely detailed distinctions.

Selections in this book reflect a long commitment to my home territory in the Pacific Northwest, to knowing the land I live in, my gardens and hiking and camping within a day's drive. Of course I continue on family visits and vacations everywhere. Of the more than eight hundred species of butterfly known in the United States and Canada, a good sampling of a hundred and fifty is included here.

Public lands, owned and managed for the common good by federal, state, and local governments, hold much of the wild and semi-wild habitat where I found and photographed most of the butterflies in this book. Such lands are named throughout the book, to honor their value.

Chapters begin with a gallery of photographs, followed by illustrated discussions of various aspects of butterfly biology.

As I saw different sizes, shapes, and color patterns, curiosity drew me into first learning the species names, and then their relations to each other, and a gradual understanding of where to find them. Lepidopterists have organized our North American butterflies into half a dozen families, with many more subfamilies, which are introduced here. Photographs sample the variety of species, and some interesting characteristics are described.

Markings by which we can know one species from another are discussed and illustrated in field guides. Often we don't get long to look at a wild butterfly, so here are some suggestions on how to study and observe.

Our beloved Monarchs are shown in many aspects of their lives, from coast to coast, and visited in their overwintering sanctuaries in Mexico.

Birds, spiders, mice, and many more animals can see butterflies as food to eat. A chapter on threats and butterfly death explores these vulnerabilities and their defenses.

Many aspects of reproduction are shown, including courtship, mating, eggs, caterpillars, and freshly emerged adults.

So where do we find these fascinating butterflies? Almost anywhere that has the needed plants—larval hosts and nectar sources. Home gardens, semi-wild countrysides, and the back country. There are "hot spots" of abundance. Especially attractive flowers for nectaring are shown—and your area will have its own. Other photographs show butterflies along with the habitats where they are found.

Next are more butterflies, with a focus on people studying and enjoying them.

We close with a discussion of conservation issues, and with the humanity of our interest in butterflies, with photos of children butterflying.

An appendix includes a list of resources—books, internet, people—for learning more and for finding promising places to look; plus guidelines for making good photographs. A few technical terms are used, so there's a brief glossary. Scientific names, locations, and dates for all the photos are listed in the section headed "Species, Locations, and Dates."

Acknowledgments

In my wife Alexandra, I found a marvelous hiking and camping partner who encourages us to go out and then puts up with interminable cycles of searching and exposing propelled by the photographer's mantra, "Just one more." She made the vast preponderance of this work possible. My parents, Janet Vanderwalker Myers and Lawrence S. Myers, Jr., raised me steeped in art, science, and nature. An extended family with brothers Fred and Lee and my parents, joined by auxiliary parents Charles and Norma McKinney and sister Kathy for regular camping and hiking, made outdoors time comfortable, rewarding, and frequent, just the natural place to be. My first wife, Elaine, took me to near-wilderness to live and experience daily.

So many more people's energies flowed through me to accomplish this work: My photography instructor at the Associated Students of the University of California, Berkeley, Dave Bohn, long ago inspired making only the finest photographs and books, and recently encouraged me in this book. While I studied with him, he set an example with his *Glacier Bay: The Land and the Silence* (Sierra Club, 1967). Two years later Roger Minick published *Delta West: The Land and People of the Sacramento–San Joaquin Delta* (Scrimshaw Press, 1969). I was hooked—that's what I wanted to do.

Robert Michael Pyle, the butterfly evangelist and mentor who rekindled my childhood interest, helped me, for decades, to learn to find and recognize butterflies, to handle specimens, offered presentations to give, and invited me to participate in his Northwest field guides. He introduced me to the community of lepidopterists.

At the Northwest Lepidopterists Society workshops through the years, lepidopterists Andrew Brower, Paul Hammond, John Hinchliff, Caitlin LaBar, David McCorkle, Jeffrey Miller, Jonathan Pelham, Harold Rice, Dana Ross, Jon Shepard, Ray Stanford, Andrew Warren, and dozens more with unfailing and patient generosity helped me learn what I was photographing—and enthused about the value of my photographs. The meetings are hosted by the Oregon State Arthropod Collection, director David Maddison and curator Christopher Marshall. Many field guide authors and the Butterflies of America website inspired and supported my efforts.

Neil Maine and the North Coast Land Conservancy commissioned a guide to the butterflies of Oregon's north coast, supporting my first professional activity in this field.

Nathan Lyons, Robert Adams, and LightBox Photographic Gallery members gave me encouragement in the photographic work. Michael Wyman kept me wisely equipped and supplied.

Editor Jean Thomson Black understood my vision, had faith in it, and knew how to complete it. Phillip King was wonderful to work with, making sure the text is clean and flows well.

Two reviewers made the book much better by thorough, detailed reading and suggestions. Butterfly names were corrected, improvements in the science suggested, and problems with the writing noted.

Thanks also to the establishers, rangers, and administrators of public lands, federal, state, and provincial, which provide habitat for so many populations of butterflies, as well as the holders of private lands and conservancies who maintain habitat and allow access.

Thank you all. It's a good story we've told.

Wings in the Light

Butterflies

WHY BUTTERFLIES?

Every path into wild nature enriches us. Butterflies, birds, wildflowers, mushrooms, forests, tidepools, fishing, and hunting all provide great journeys beyond civilization. My
fascination with butterflies began when I had filled my head with watching the birds
and progressions of the seasons at my home. I lived in a clearing, with lightly managed
forestry on three sides and a casual cow pasture on the fourth. In Washington's Wahkiakum County, on the Columbia River near the Pacific, wildness came to me as I went
between house, shop, and garden. When I was ready to extend my curiosity, the great
butterfly evangelist Bob Pyle had moved into the neighborhood and become a friend. I
quickly found his favorite subjects attractive and approachable. Then I joined a group
of Pacific Northwest lepidopterists for annual meetings, and discovered that butterfly
study is accessible. With their support I learn what I'm seeing and why. Butterfly and
moth scientists welcome and support the work of amateurs, who make many important
contributions.

As people study biogeography and changes due to climate warming, it turns out that butterflies are especially useful study subjects. Most species are very particular about where and how they live, in association with what plants, and when those plants appear. Temperature, sunlight, humidity, and wind conditions affect them greatly. It turns out that research into these details and patterns is especially practical with butterflies, and the results are useful for understanding how to maintain habitat for them and many other kinds of organisms facing pressure from human activities, including from climate change. When we take good care of habitat for butterflies, we provide for many other animals and plants too. Shifts in the timing and location of butterfly appearances are good indicators of weather and climate changes.

There is a challenging but not overwhelming number of butterflies. Eight hundred or more species occur in the United States and Canada, with several times that many subspecies. Including all of the variable appearances and populations within subspecies, there's enough to reward a lifetime of study: we're not going to run out of interesting questions and discoveries.

Western Tiger Swallowtail, Wahkiakum County, Washington

California Tortoiseshell, Cascade-Siskiyou National Monument, Oregon

Woodland Skipper, Wahkiakum County, Washington

ADDITIONAL INSPIRATIONS

Love of light and lens. Light carries energy and liveliness. Detail offers us ways to engage. Form and design organize our attention. We can soften our perception, for the emergence of connections to our internal being and to the vast web of existence. If there's a secret here, it's in slow seeing. In slow seeing with quick eyes.

Public lands and private habitat. Places that support wild and semi-wild habitat, and welcome roaming by naturalists, are much appreciated.

Mormon Metalmark, Fisher Hill Unit, Washington
Department of Fish and Wildlife

WHAT ABOUT THE GORGEOUS BUTTERFLIES OF THE TROPICS?

Upon learning that I'm a butterfly photographer, people often ask if I photograph in the tropics. I love the tropical butterflies as much as anyone. They're spectacular, interesting and important to study, and get well-deserved media attention. I hope that's a useful contribution to preservation of butterflies. The role I have chosen is to work with the temperate-zone butterflies of my own home territory. I do include the overwintering Monarchs in Mexico, a population shared with us in the north. I want to help us be fascinated by the life here in our own surroundings, our backyards, the hills we hike, and the lands we consider for our city and industrial developments, and for agriculture, forestry, and mining.

Obsessive knowledge and dedicated work are required to maintain the habitat and climatic conditions for butterflies to thrive. Places near our homes are satisfying to work to maintain the fullness of natural phenomena. For humanity, what's at stake is whether we use the grand, wild resource of nature to enrich our lives, and whether we hand it on to forthcoming generations for their enrichment.

Mourning Cloak, Ontario State Park, Oregon

Fritillary, very old and worn; Steens Mountain
Recreation Lands, Oregon

Chalcedona Checkerspot, Northern Sonoma
or Napa County, California

Although I have chased butterflies throughout North America, most of my work and results are near my home, especially in Oregon and Washington.

Silvery Blue, Gifford Pinchot National Forest, Washington. This fellow, a male, did a 360-degree turn for me in about twenty seconds, on a vertical bank in good light.

(left) Anna's Blue, Willamette National Forest, Oregon

(above) Echo Azure, on kayak guide Ginni Callahan; Julia Butler Hansen National Wildlife Refuge for the Columbian White-tailed Deer, Washington

(facing) Common Ringlet, Deschutes River National Recreation Lands, Oregon

Boisduval's Blue, Sawtooth Wilderness, Sawtooth National Forest, Idaho

Aphrodite Fritillary, Allegheny National Forest, Pennsylvania

Families of Butterflies

FAMILIES OF BUTTERFLIES: TAXONOMY

Butterflies have evolved through tens of millions of years, splitting into many different species to thrive on various plants and habitats. Lepidopterists—butterfly scientists—organize these by similarities and differences. The key concept is the species. Organisms within a species are generally able to reproduce effectively with each other, but not with members of other species.

Many species of butterflies live in regional populations that don't mix much beyond their boundaries. These can have unique appearance and behavior slightly different from other populations of the same species. Or sometimes there's a long clinal transition from one type to another, with plenty of variety in each locale along the way. If differences are significant enough, the species is split into subspecies, each representing valuable genetic information of adaptation to their local habitat. Closely related species are grouped into a genus; a number of these genera are gathered into families; and additional terms are used for varying degrees of relatedness. Subfamilies are used in the following pages to organize the variety of North American butterflies. Lepidopterists have hearty debates about taxonomic issues, and every source you consult will use at least a few different names and arrangements.

August in the Idaho Rockies found most of the woods dry, but this moist meadow was full of life. I call it Twenty-Minute Meadow, because that's all it took to photograph six species: (facing) Greenish Blue, Western Branded Skipper, Mormon Fritillary, Field Crescent, the same Crescent, Sonora Skipper, Dark Wood Nymph; Challis National Forest, Idaho.

Northern Cloudywing, Oregon

Long-tailed Skipper, North Carolina

Horace's Duskywing, Virginia

Northern White-Skipper, Washington

Common Checkered-Skipper or White Checkered-Skipper, Texas. Whether a Checkered-Skipper is Common or White can be determined only by microscopic dissection.

Spread-wing Skippers often perch with their wings open flat. There are many kinds of tan and brown ones, with maddeningly subtle differences.

Propertius Duskywing, Oregon

Common Checkered-Skipper, Bureau of Land Management, Oregon. We know it's not a White Checkered-Skipper because it was seen far from the latter's range.

Persius Duskywing or Propertius Duskywing, Deschutes National Forest, Oregon

FOLDED-WING (GRASS) SKIPPERS: SUBFAMILIES HETEROPTINAE AND HESPERIINAE

Arctic Skipper, Oregon

Woodland Skipper, Oregon

Juba Skipper, Idaho

Fiery Skipper, Texas

Peck's Skipper, Idaho

Common Roadside-Skipper, Oregon

Folded-wing Skippers often perch with their hindwings open flat and forewings half closed, or sometimes with all wings closed overhead. There are many kinds of orange, tan, and brown ones, with challenging, subtle differences.

Rural Skipper, Wild Rogue Wilderness, Oregon

Least Skipper, New York

Woodland Skipper, Oregon

Mountain Parnassian, Idaho

Clodius Parnassian, Oregon

Anise Swallowtail, Oregon

Indra Swallowtail, Oregon

Parnassians are tail-less members of the swallowtail family, remarkable for substantial clear areas on their wings.

Large, colorful swallowtails come in bold patterns of yellow, white, and black, with exciting touches of blue, red, and orange.

Western Tiger Swallowtail, Cape Disappointment State Park, Washington

Anise Swallowtail, Washington

Western Tiger Swallowtail, Malheur National
Wildlife Refuge, Oregon

SWALLOWTAILS OF THE EAST: SUBFAMILY PAPILIONINAE

Female Eastern Tiger Swallowtails can be either yellow or black, as shown here. The black ones are thought to be mimics of the distasteful Pipevine Swallowtail.

Eastern Tiger Swallowtail, yellow female; Virginia

Eastern Tiger Swallowtail, black female; Virginia

Palamedes Swallowtail, Georgia

Eastern Tiger Swallowtail, Scott's Run Nature
Preserve, Fairfax County, Virginia

Eastern Giant Swallowtail, Caley Reservation, Lorain County Metro Parks, Ohio

Black Swallowtail ovipositing, National Butterfly Center gardens, Texas

SULPHURS: SUBFAMILY COLIADINAE

Clouded Sulphur (left), Orange Sulphur
(right), Pennsylvania

Clouded Sulphur, white form; Wyoming

Cloudless Sulphur, Arizona

Western Sulphur, Oregon

Medium-small sulphurs are usually yellow, though some have orange or green tints, and there are white forms of several species. These normally perch with their wings folded together, showing the ventral (underside) surfaces of the wings; we usually see the dorsal (upper) surfaces only when they are in flight. The Dainty Sulphur is tiny.

Pink-edged Sulphur, Réserve Faunique
La Vérendrye, Quebec

Orange Sulphur, South Dakota

Dainty Sulphur, National Butterfly Center gardens, Texas

Cabbage White, Oregon

Large Marble, Oregon

Great Southern White, Florida

Sara's Orangetip, Oregon

Pine White, Idaho

Western White, Idaho

Cabbage Whites cheer our yards with their sunny flight, while their caterpillars compete effectively with us for our homegrown kale and broccoli. Sara's Orangetip is curving her abdomen down and forward, likely to lay an egg on this mustard family plant.

Becker's White, Price Canyon Recreation Area,
Utah

Great Southern White, Spanish Harbor Wayside Park, Florida

Arctic White, Porcupine Creek State Recreation Site, Alaska

Purplish Copper, male; Oregon

Lustrous Copper, California

Harvester, Virginia

Blue Copper, Idaho

Edith's Copper, Oregon

American Copper, Virginia

Coppery or orange colors abound. Some of the brown females look enough like brown female blues to keep us on our toes. The Blue Copper rebels, having bright blue males.

Purplish Copper, female; Deschutes River National Recreation Lands, Oregon

Tailed Copper and Fred Myers, Ashley National Forest, Utah

Brown Elfin, Oregon

Dusky-blue Groundstreak, Texas

Western Pine Elfin, Oregon

Coral Hairstreak, Idaho

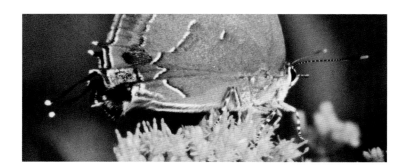

White-M Hairstreak, Virginia

Eastern Tailed-Blue, Illinois. Closely related Tailed-Blues also wiggle their wings, as shown in this pair of photos.

Many of these have bright eyespots and hair-thin tails, creating a false head and antennae impression. The wings are rubbed alternately up and down, wiggling the tails. Birds and jumping spiders are effectively diverted to attack these false heads, leaving the escaping butterfly with only nuisance damage.

Juniper Hairstreak with Ann Musché, Steens Mountain Recreation Lands, Oregon

Gray Hairstreak,
Metolius Preserve,
Deschutes Land
Trust, Oregon

Golden Hair-
streak, Rogue
River National
Forest, Oregon

Echo Azures, and Western Tailed-Blue (right); Oregon

Echo Azure, female, flying; Oregon

Melissa Blue, Idaho

Marine Blue, Arizona

Silvery Blue, Oregon

The blues are thumbnail size, quite colorful when the light catches them just right—especially when flying or congregating. For identifying them, it's almost always essential to see the ventral surfaces on the underside of the wing, and often invaluable to see the dorsal, too.

Melissa Blue, Black Canyon of the Gunnison National Park, Colorado

Blue, Muncho Lake Provincial Park, British Columbia

Boisduval's Blue, Challis National Forest, Idaho

Male, dorsal   Male, ventral   Female, dorsal   Female, ventral

All are Greenish Blues from the same meadow, in Oregon

In many species of blues the females are brown. Shown here are males and females of the Greenish Blue. For many butterflies, males and females are different enough that we learn two different coloration patterns, along with some features they share.

Northern Blue, Congdon Government Campground, Yukon

METALMARKS: SUBFAMILY RIODININAE

Unidentified Calephelis, Mexico; similar-looking individuals occur in the United States

Fatal Calephelis, California

Fatal Calephelis, Texas

A few members of this large tropical and subtropical family of highly varied and bright coloration appear in our northern latitudes. Our Metalmarks are small butterflies, some of which have rows of bright metallic looking spots.

Mormon Metalmark, Fisher Hill Unit, Washington Department of Fish and Wildlife

American Snout, Arizona

American Snout, North Carolina

Snouts have elongated mouth parts (palpi), making a unique appearance.

American Snout on the arm of the photographer, Frederick L. Myers;
Eno River State Park, North Carolina. The end of the proboscis shows
the sensilla—a sense organ.

Queen, Texas

Queen, Texas

Monarch, Mexico

Monarch, Virginia

Milkweed butterflies are a worldwide tropical and subtropical family, whose larvae feed on milkweeds. Our North American representatives—Monarchs, Queens, and Soldiers—are orange and black. These large butterflies are powerful fliers—especially the migrating Monarchs. Around the world, other danaids vary greatly in appearance and size, especially as they participate in mimicry complexes.

Queen, National Butterfly Center gardens, Texas

LONGWINGS AND FRITILLARIES: SUBFAMILY HELICONIINAE

Bright eye-catchers, medium large. The dorsal (upper) surfaces are very similar on many species, making it especially important to see and study the ventral (underside). A photograph showing only the dorsal wing surfaces can often not be identified exactly.

Gulf Fritillary, Florida

Zebra Longwing, Florida

Variegated Fritillary, Colorado

Fritillary, likely Zerene, Oregon

Hydaspe Fritillary, Oregon

Meadow Fritillary, New York

Western Meadow Fritillary, Washington

Zebra Longwing, Bahia
Honda State Park, Florida

Variegated Fritillary, Big
Bend National Park, Texas

Great-Spangled Fritillary,
Ochoco National Forest,
Oregon

TRUE ADMIRALS, SISTERS, AND EMPERORS: SUBFAMILIES LIMENITIDINAE AND APATURINAE

A habit of gliding between wing flaps helps us recognize these medium-large and colorful butterflies.

Lorquin's Admiral, California

White Admiral, Alaska

California Sister, California

Hackberry Emperor, Illinois

California Sister, Wild Rogue
Wilderness, Oregon

Tawny Emperor, National Butterfly Center
gardens, Texas

Lorquin's Admiral, Ecola Creek Forest Reserve, City of Cannon Beach, Oregon. Males take high, commanding perches from which to defend territory.

LADIES AND ADMIRABLE: SUBFAMILY NYMPHALINAE, GENUS *VANESSA*

The genus *Vanessa* includes four species common in North American yards and gardens, and a fifth in Hawaii. Red Admirables, often called Red Admirals, are in *Vanessa*, not the genus *Limenitis* of the true admirals.

West Coast Lady, Oregon

American Lady, Oregon

Red Admirable, Oregon

Red Admirable, Wallowa Lake State Park, Oregon

PAINTED LADIES

Painted Ladies are one of our most common butterflies—in some years. They cannot survive the cold winters in most of the United States, not in any life stage—egg, caterpillar, chrysalis, or adult. Instead, a core population overwinters in the southwestern deserts and in northern Mexico. When a winter is wet so thistles and other host plants for caterpillars are abundant, great numbers are produced. These migrate north and fan out, covering the country from shore to shore. Along the way they find host plants, lay eggs, and become even more numerous. During the most intense irruptions, clouds of Painted Ladies have interfered with traffic. In drier years, their numbers are scarce. It has long been believed that there's no return flight to the south, but recently some evidence has been found of high-altitude returns.

Schoolchildren often have the marvelous experience of raising Painted Lady caterpillars and releasing the adult butterflies. Generally biologists are very cautious about raising and releasing anything outside of its home territory: Such releases confound studies of what appears when and where, as well as introducing genes that may be adapted to

Painted Ladies, Wahkiakum County, Washington

another area's different conditions, and can also spread diseases. When a Monarch is observed today, for instance, one never knows for sure if it got there on its own or from a wedding release. The best practice is to find local butterfly eggs and raise them.

A great Painted Lady irruption in 1995 was so rewarding to my camera that butterflies became a major aspect of my work. I would often have five or ten at a time in one little patch of asters. They were so engrossed in nectaring that they allowed me to lie among them with my macro lens. All the photos on these three pages are from that season.

Painted Ladies, Wahkiakum County, Washington

CLOAK AND TORTOISESHELLS: SUBFAMILY NYMPHALINAE

In colder areas, the first butterflies seen in late winter and early spring often include Cloaks or Tortoiseshells, which hibernate through the winter as adults.

Mourning Cloak, Oregon

California Tortoiseshell, Washington

Milbert's Tortoiseshell, Idaho

Milbert's Tortoiseshell, same individual

Milbert's Tortoiseshell, Bitterroot National Forest,
Montana

ANGLEWINGS (COMMAS): SUBFAMILY NYMPHALINAE

The genus *Polygonia* is named for the many sides on the jagged wings. With falcate (hooked) wingtips, they are fast and direct in flight. Many have "startle camouflage" with dull, striated ventral surfaces that conceal them when perched on limbs, contrasting with their bright orange dorsal surfaces. Some references call this group Commas.

Question Mark, Texas

Comma Anglewing, North Carolina

Green Anglewing, Oregon

Hoary (Zephyr) Angle-
wing, Department of Fish
and Wildlife, Oregon

BUCKEYES: SUBFAMILY NYMPHALINAE

Spectacular eyespots and other colorful markings adorn these brown, medium-sized butterflies.

Common Buckeye, Florida

Common Buckeye, same individual

Common Buckeye, California

Mangrove Buckeye, Florida

Mangrove Buckeye, Ding Darling National Wildlife
Refuge, Florida

CHECKERSPOTS AND PATCHES: SUBFAMILY NYMPHALINAE

Red, orange, and black checkerboards, or sometimes large patches or swaths of color, mark these medium-small butterflies. Details of complex ventral-surface markings are often critical for identification.

Edith's Checkerspot, Oregon

Snowberry Checkerspot, Oregon

Sagebrush Checkerspot, Oregon

Leanira Checkerspot, Oregon

Bordered Patch, Texas

Anicia Checkerspot on Thea Linnea
Pyle, Wenatchee National Forest,
Washington. This Anicia was gently
netted, placed on Thea, and chose
to stay. Sometimes they land on us
of their own accord.

Chalcedona Checkerspot, Cascade-Siskiyou National Monument, Oregon

Second view, same individual

CRESCENTS: SUBFAMILY NYMPHALINAE

Small, bright butterflies in orange and black. Near the ventral hindwing margin, there's usually a prominent crescent, often in a softly defined dark area. The dorsal hindwing margins tend to have a row of subtle crescents.

Mylitta Crescent, female; Idaho

Mylitta Crescent, male; Oregon

Field Crescent, Oregon

Pale Crescent, Oregon

Texan Crescent, Texas

Mylitta Crescent, probably bird-struck; Eagle Cap
Wilderness, Wallowa-Whitman National Forest,
Oregon

SATYRS: SUBFAMILY SATYRINAE

Our North American Satyrs are usually medium-small brown or orangeish butterflies. Their flight is often soft and gentle. Many have a habit of gliding with their wings closed overhead, likely as a way of maneuvering through the grasses and sedges on which their larvae live.

Common Ringlet, New York

Northern Pearly-eye, New York

Common Alpine, British Columbia

Dark Wood Nymph, Colorado

Chryxus Arctic, Idaho

Great Arctic, Oregon

Great Arctic, Deschutes National Forest, Oregon

Common Ringlet, New York

Common Wood Nymph, Coeur d'Alène National Forest, Idaho

Observing Field Marks

Butterfly field guides are designed to help us identify what we see. They cover the characteristics of species and the distinctions between them, with good illustrations and descriptions. General appearances can lead you to the right family, and specific markings to the species. When seeing a butterfly, start by noticing the overall intuitive effect, the general size, shape, color, behavior, and surroundings, so you will turn to the right chapter in the guides. Watch for any diagnostic details you can catch.

Notice how a butterfly flies—fast and powerfully? Slowly and flutteringly? Does it glide—with wings out flat? Folded overhead? While we find many butterflies on flowers, others are characteristically found perching on trees, shrubs, or herbs. Many may also be found basking on rocks, logs, or the ground. Every detail helps with identification.

When I show a butterfly photograph to more experienced lepidopterists, their first question is about what plants were present and absent. Flowers are great for finding nectaring butterflies, but for presence of a species what matters is the host plants for caterpillars. Treasure your botanist friends, and dig into their books.

You may find it easiest to start with a regional field guide showing only species in your area. Sort through two hundred species instead of eight hundred. Range maps and flight season descriptions help us focus on the most likely possibilities. For sure, some individuals show up in unusual places and times, making exciting discoveries. Such are easier to recognize and trust as one becomes experienced with what's common.

In my experience, knowing what characteristics to look for, from books and experienced butterfliers, helps me see more accurately when in the field. It's back and forth between the books and observing, between studying what markings are useful and looking for them. Over many seasons, both my knowledge and my seeing get better.

When photographing for identification, catch the body and every wing surface that you can. Details in a blurry picture or a view partly obstructed by leaves or sticks can still be valuable for identification. Often I can record useful amounts of upper and lower wing surfaces in the same photo, which is especially valuable when there are several individuals around. Otherwise one might not be sure if one photo of upper surfaces and another of the under surfaces are on the same individual. It's not unusual to find several species flying together.

Oregon Swallowtail, Oregon, two views. By lying on the ground I saw the ventral wing surfaces and the sides of the body.

Acmon or Lupine Blue, Oregon

Western Tiger
Swallowtail

Anise Swallowtail

Western Meadow Fritillary

Painted Lady

West Coast Lady

SCINTILLAE AND SILVERED SPOTS

The presence or absence of scintillae—shiny spots near the underside hindwing margin—is a valuable fieldmark. These sparkle with light at just the right angle. In North America these appear mainly on some blues and a few satyrs.

Silvered spots appear on the underside hindwings of many *Speyeria* fritillaries. Always a good clue, they're rarely decisive, with many populations having both silvered and unsilvered individuals. When they catch the sun, these spots are startlingly bright silver.

Northern Blue, Wyoming

Western Pygmy Blue, Oregon

Nabokov's Satyr, Arizona

Coronis Fritillary, California

Cassius Blue, Bahia Honda State Park, Florida

Monarch, Oregon

Monarchs

Monarchs are surely our best known, most widely appreciated butterflies throughout the United States and southern Canada. Their powerful flight and gliding on immense orange wings with black lines and white spots carries a regal presence.

Fertile female Monarchs have evolved a way to protect their caterpillars and future adults from predators such as birds. They lay their eggs only on milkweed plants, which produce toxins making them inedible to most creatures. Monarch caterpillars have evolved to both tolerate the toxins and incorporate them into their bodies, which then are inedible to many predators. These toxins remain in the adult butterflies and give them substantial protection from hungry birds.

MILKWEED, CATERPILLAR, AND CHRYSALIS

A nectaring Monarch will make good use of the food energy from flowers in our yards and countrysides, but her caterpillars will require milkweeds that hopefully can be found nearby.

 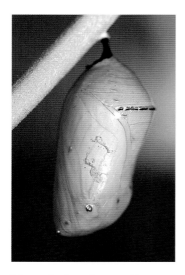  

Monarch caterpillar on milkweed, Oregon. When all goes well on the milkweed, the caterpillar will crawl to a plant nearby, make a chrysalis, and later emerge as a butterfly.

Monarch chrysalis raised by Dr. David James, associate professor of entomology at Washington State University. His doctoral thesis was on Monarch biology. The dark veining pattern of the wings forming inside is faintly visible on the left of this view. (Photo by David G. James)

Not a Monarch: an Owl Butterfly that has just eclosed, or emerged, from its chrysalis. Wings are soft and wrinkled as they grow inside the chrysalis. This butterfly has begun pumping fluid into its wing veins to stretch them flat. (Photo made in a butterfly house in Ecuador)

Once the wings have been stretched flat, the butterfly waits for them to dry and stiffen. This one is a Monarch, and soon it will be ready for flight. (Photo made in a butterfly house in Ecuador)

Monarch, Virginia

In northern milkweeds, Monarchs have found a food source that not many other creatures compete for. But no life stage of the Monarch can survive the cold, snowy winters. Monarchs from most of the interior United States, the eastern seaboard, and southern Canada have solved this problem with a migration cycle in which they spend summer in the north and winter in tropical mountains. The final summer generation of Monarchs is specialized for long life and long flight. These fly to south-central Mexico, a journey of up to two and a half thousand miles.

Certain peaks of the Trans-Mexican Volcanic Belt between Mexico City and Morelia provide just the low temperatures—slightly above freezing—that the butterflies require to make it through the winter. They need cool weather to slow their metabolism so they won't run out of stored fat energy, but without a killing freeze. They also need the oyamel fir trees that grow here, which provide them with sheltered high-density roosts.

Then in March they head back north, flying until they find milkweeds to lay eggs on, in the southern tier of the United States. The generation from these eggs, and one or two more, are specialized for much greater reproduction, at the cost of shorter lives. They continue northward, fanning out eastward and westward, reaching well into Canada to fill the countryside with butterflies. Then as autumn cools, the cycle turns southward again, with another long-distance generation heading to Mexico.

How do we know the migrations? Researchers in the north catch Monarchs to put tiny numbered tags on them. Then people search the overwintering groves, and occasionally find a tag, revealing where the butterfly came from.

A tagged Monarch, at Santuario de la Mariposa Monarca el Rosario, near Angangueo, Michoacán: A Mexican steward of the Monarch groves shows a tag recently recovered, while photographer Steve Trimble documents the occasion. Researcher David Marriott (left, white T-shirt) will record and analyze the data.

ENTERING THE SANCTUARY

The Mexican migration phenomenon of North American Monarchs is endangered, threatened by several factors, including, in the north, fewer milkweeds to host their caterpillars, and also agricultural insecticides. Milkweeds are rarely tolerated in modern agricultural, residential, commercial, and industrial areas. Planting of crops genetically modified to tolerate herbicides is followed by heavy application of herbicides to kill everything else, incidentally preventing milkweeds from growing along nearby fence lines, unplanted areas, and roadsides. In Mexico, illegal logging in their wintering groves destroys some of the Monarchs' options for shelter, while climate change makes the timing and details of temperature and moisture unreliable at their overwintering groves.

Unfortunately these problems have combined to make Monarch populations insecure and to threaten this unique insect migration. An agreement reached in 2014 between the presidents of Mexico and the United States and the prime minister of Canada to "establish a working group to ensure the conservation of the monarch butterfly" may help.

Entrance to Santuario de la Mariposa Monarca el Rosario, in Michoacán, Mexico

Vendors at Santuario Mariposa Monarca Sierra Chincua, Michoacán

Following our guide into Santuario Chincua

Monarchs in flight, as we hike into Santuario Chincua

Monarchs basking in the sun, Santuario Chincua

More than five thousand Monarchs are in this photograph, roosting in clusters at Santuario Chincua

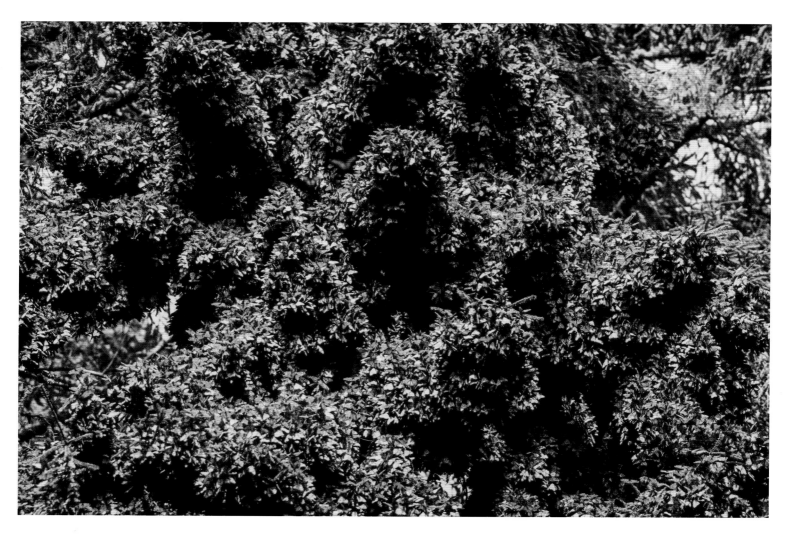

Monarchs roosting on oyamel fir trees, Santuario Chincua

DRINKING WATER

Overwintering Monarchs in their forest reserves of central Mexico need water regularly. A steward from the nearby community waters a patch of bare dirt and a grass lawn, supporting both his family and the butterflies with donations from visitors. Visitors walk gently among masses of calm butterflies.

At these grounds a magical spirit appears: Robert Michael Pyle. Bob was part of the small group that negotiated the founding of the Monarch preserves in the state of Michoacán, Mexico. He is an evangelist for engaging butterflies and natural history in general.

Monarchs drinking water at Santuario de la Mariposa Monarca el Rosario

Robert Michael Pyle with Monarchs at Santuario Rosario

Monarchs massed on wet ground at Santuario Rosario

PREDATION AND MATING IN THE SANCTUARY

Monarchs' toxin shield against bird appetites is good, but not perfect. Three birds are able to feast on the massed overwintering Monarchs in Mexico—Black-backed Orioles, Scott's Orioles, and Black-headed Grosbeaks.

Orioles slice the Monarch abdomens open and strip out the contents, avoiding the higher concentration of toxins in the shell. Grosbeaks are able to eat the entire abdomen. Sometimes the thoracic muscles are also eaten. Walking through the groves in some places the forest floor will be thick with wings and partial bodies.

Before flying north, some Monarchs mate to fertilize their eggs; others wait until they are near or at the southernmost milkweed patches where they'll lay eggs.

Dead Monarchs, their abdomens eaten by birds; Santuario Chincua

Monarchs mating, Santuario Chincua

PRIVILEGE: WHO CAN GO BEYOND HERE?

Santuario Sierra Chincua in Michoacán protects a major overwintering grove. Mexican schoolchildren are bused in so they can learn about and enjoy the Monarchs in the sanctuary. The schoolchildren and most other visitors were allowed to go only this far, where, indeed, many butterflies are visible. But even better viewing lies beyond. British and Japanese film crews were allowed farther in. Because our party of tourists was led by two scientists involved in the establishment and current research of the sanctuaries, we were also allowed to go farther than the children.

Was this fair? Foreign tourists allowed better access than local schoolchildren? Some of us are the products of privilege: well connected and wealthy. We have no choice about that. But we do have a choice of how to use that privilege. Consider privilege to be a responsibility, to be used gently and generously, with care for the common good.

Mexican schoolchildren
and Monarchs,
Santuario Chincua

WESTERN POPULATIONS

Monarchs from much of the West Coast depend on overwintering in California coastal groves, a few minutes' walk from the ocean beach. Such places are also beloved by people for our own activities, and especially for constructing buildings and parking lots, which replace the trees and are useless to butterflies.

FLORIDA AND CUBA

Southern Florida hosts a non-migratory population, joined by migrants from the Atlantic states. Some research suggests that Monarchs from the central and eastern states fly to Cuba, and likely do not contribute to a northward return.

Monarch, Key West Art and Historical Society's Custom House Museum, Florida

Goleta Butterfly Grove, near Santa Barbara, California. A campaign from 2002 to 2005 developed a coalition of governments, a housing developer, and private and public donors that resulted in the formation of the Goleta Butterfly Grove on the Sperling Preserve. Interestingly, the native pines that once grew here are largely gone, but the butterflies have adapted to the eucalyptus trees that were introduced.

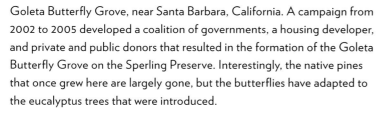

Monarchs at Goleta Butterfly Grove, California

MASK OF THE MONARCH

I photographed monarchs on their roosts in trees and flying through shafts of forest sunlight and blue sky. I also looked down to the forest floor where eucalyptus leaves, seed pods, and a few unlucky butterflies had died and fallen. Mice were glad to eat the bodies, leaving wings loose to blow about among the eucalyptus debris.

This wintering grove has been protected from development. The Monarch population and migrations remain threatened by agriculture and development in the United States and Mexico, and by climate change.

A mirror image splice of my forest-floor photograph revealed this ancient American Monarch mask watching us. Will we continue to leave a place for their lives, or will we take every last piece of coastal habitat for our own purposes?

Monarch wings and Eucalyptus debris, found on the forest floor; Sperling Preserve, California

Threats and Death

PREDATION

A butterfly lives its life to survive and reproduce. Hungry spiders, birds, lizards, and insects have other ideas, seeing them as tasty morsels of fat and protein. Most eggs and caterpillars, and many adults, are eaten.

Four Tawny Emperors dine on rotting banana bait at the National Butterfly Garden in Mission, Texas. Five seconds later one of them became a lizard's snack.

Tawny Emperors, male left, female right; Texas

Tawny Emperor and lizard, National Butterfly
Center gardens, Texas

TAILS AND SPOTS

When birds strike butterflies, sometimes the butterflies escape with acceptable wing damage, their bodies spared. The tear can even have the shape of a bird beak bite. Butterflies often manage quite well with big gaps in their wings. Eyespots and tails on the wings can serve as diversionary targets for birds, as sacrificable areas.

I have watched a Barn Swallow pursue a Western Tiger Swallowtail, the two zigzagging in flight until the bird gave up.

Common Buckeye, North Carolina

Eastern Tiger Swallowtail, North Carolina

Arctic Fritillary, Gifford Pinchot National Forest, Washington

Red-spotted Purple, Pennsylvania

Pale Tiger Swallowtail, Ochoco National Forest, Oregon

SPIDERS

Some spiders catch butterflies and other insects in webs. Others wait in flowers to attack, kill, and eat nectarers.

Sulphur in a web, with spider; Washington

Likely a Western Branded Skipper, caught by a spider; Ochoco National Forest, Oregon

Woodland Skipper caught by a spider, Cascade-Siskiyou National Forest, Oregon

I photographed this Yehl Skipper in a flower several times, moving closer and closer, wondering why I hadn't scared it off. Only then did I see the spider.

Yehl Skipper caught by a spider, Point Lookout State Park, Maryland

Common Ringlet caught by a spider, Deschutes
River National Recreation Lands, Oregon

California Tortoiseshell caught by a spider,
Cascade-Siskiyou National Monument, Oregon

MORE PREDATORS

One day in my garden a bald-faced hornet struck a Purplish Copper butterfly midair, took it to the ground for a moment, then flew its catch to a high alder branch. The hornet may have consumed the butterfly, or taken it to its nest to feed its own larvae.

Fritillary with bee and chalcid wasp, Idaho

Detail from the left forewing

Tiny wasps often parasitize caterpillars: they lay their wasp eggs in the butterfly egg or caterpillar, then the wasp larva lives in the caterpillar and devours it from the inside. I noticed this chalcid wasp on the left forewing of a fritillary. Don't know what it was doing there.

Female Western Branded Skipper caught by an ambush bug, Bureau of Land Management, Winnemucca District, Nevada. It's easy to see how this ambush bug is all but invisible on the flower.

CAMOUFLAGE

Some butterflies use camouflage to avoid being noticed and eaten. Many show bright wing surfaces when flying, and then upon landing fold their wings and become quite inconspicuous. Imagine following a bright orange flying object, and then trying to find the dull gray-brown stationary one it becomes. Or closing in on the plain brown insect, and suddenly seeing it fly away in a flash of orange. It's called "startle camouflage."

Milbert's Tortoiseshell, the same individual, two views; Malheur National Wildlife Refuge, Oregon

Hoary (Zephyr) Anglewing, Challis National
Forest, Idaho

Hoary (Zephyr) Anglewing, Deschutes National
Forest, Oregon

Ornythion Swallowtail, Texas

Monarch, Virginia

MIMICRY

Mimicry is a way of avoiding being eaten by looking like something inedible, rather than by avoiding being noticed. The Ornythion caterpillar looks like an unappetizing bird dropping.

The distasteful Monarchs have an effective lookalike mimic in the Viceroy, which appears to be somewhat poisonous in its own way. Fortunately for butterfliers, Viceroys do have an extra black line arcing through the hindwing, visible on both dorsal and ventral surfaces. To see how the Viceroy has adapted its appearance, compare its closest relatives in this country, the true admirals—White, Weidemeier, Lorquin's, and Red-Spotted Purple—itself a member of a different mimicry complex.

Viceroy, Caley Reservation, Lorain
County Metro Parks, Ohio

RAIN

Falling raindrops are heavy enough to be dangerous to even a large butterfly. Darker skies and cooler temperatures make it harder for butterflies to fly. When rain threatens, butterflies seek shelter deep in the grass, or under leaves, or in bark crevices.

(facing) Painted Lady, in the grass, during a rain, with a detail (above) showing the red, black, and white butterfly in the center. I don't see a sight like this very often.

WORN OUT

Some escape being eaten, but eventually wear out and die of old age. Butterflies live anywhere from a few weeks to a few seasons, depending on their size and other factors. Many have only a single brood in a year, so from one year's egg to the caterpillar, chrysalis, and adult to the next year's egg is a full year, even two years for a few species.

Pine White, Bitterroot National Forest, Montana

Woodland Skipper and water strider, Fisher Hill Unit, Washington Department of Fish and Wildlife. The skipper may have died and fallen into the water, or could have contacted the surface and become trapped. Water striders will eat both live and dead insects.

Pale Tiger Swallowtail, Alum Rock Park, San Jose, California

Next Generation

MATING, EGGS, AND CATERPILLARS

Anise Swallowtail, ovipositing an orange egg (center) on garden fennel; Wahkiakum County, Washington

Anise Swallowtail caterpillar on fennel in a home garden, Wahkiakum County, Washington

Queens courting and mating, probably the same pair;
Arizona-Sonora Desert Museum, Arizona

Gulf Fritillaries, Del Rio Beach, California.
I watched probably this female lay the egg.

CELEBRATE CABBAGE WHITES

It's easy to love these busy white spirits, flitting through our yards exploring each other and our plantings. It's equally easy to be annoyed at their highly successful taking of our cabbage, kale, and broccoli.

Cabbage Whites are of European origin and thrive in habitats we have created such as home gardens and farm fields, and many wild places, wherever cabbage and mustard family or relatives are found.

This series of photographs was taken in my garden in Astoria.

Cabbage White eggs, Oregon

Cabbage White caterpillars consuming kale, Astoria, Oregon.

Cabbage Whites, mating pair on rhubarb; Astoria, Oregon

Great Basin Fritillaries, mating pair; Mount Pisgah, Ochoco National Forest, Oregon

A CONCEALED CATERPILLAR

In the woods of Shenandoah River State Park in Virginia, my brother Fred and I noticed a rolled leaf, and found that it was sheltering a Spicebush Swallowtail caterpillar. We were very careful to leave the leaf on the bush, and tucked the caterpillar back into its rolled leaf.

Spicebush Swallowtails, courting pair; Virginia

Caterpillar in rolled spicebush leaf, Virginia

The caterpillar revealed

Hydaspe Fritillary, preparing for its first flight; Ochoco National Forest, Oregon.
Its delicate hairs haven't yet been worn.

Where?

PRIMITIVE CAMPS

National Forests and Bureau of Land Management holdings give us marvelous access to wildlands. Though not pristine, due to logging, grazing, and recreation, these are often still excellent habitats, and allow plenty of wandering by the curious.

I use the camp at Mack's Canyon in Oregon as a base for spring butterflying. One common species is the Juba Skipper. Swallowtails and blues appear abundantly at the boat ramp to the Deschutes River, and in nearby hills and draws. On a climb above the campground one encounters patches of desert buckwheats, lupines, and butterflies. The Deschutes River here is within twenty-five miles of the Columbia. Wet sand attracts Indra Swallowtails. Bighorn sheep adorn the cliffs above.

Juba Skipper, Deschutes River National Recreation Lands, Oregon

Gray Hairstreak, Deschutes River National Recreation Lands, Oregon

Oregon Swallowtail (center) and Anise Swallowtail (left and right), Deschutes River National Recreation Lands, Oregon

PLANTINGS FOR BUTTERFLIES

Exuberantly overgrown gardens and yards have far more to offer butterflies than do more cleanly groomed areas. Nectar sources, perches, food plants for caterpillars, and plenty of stalks and litter for chrysalises and overwintering eggs or caterpillars are all helpful. Details vary by species.

In Fred and Cathy Myers's yard in Falls Church, Virginia, plantings from ground cover to trees, both fresh and fading, provided great habitat. More than twenty-five species of butterfly appeared.

The good nectar plants often recommended for butterfly gardening are great for bringing butterflies into our daily view. It is a joy to see what's present in the neighborhood. And in our gardens, it's easier to photograph butterflies, identify them, and share awareness of them with our friends. Though valuable, garden flowers are not enough for butterfly conservation. The real key is the mix of native nectar sources and caterpillar food plants in the countryside and wildish urban and industrial areas.

Eastern Tiger Swallowtail, Virginia

SUPPORTING HABITAT

The communities of enthusiastic citizens, government officials, and businesses that came together to keep habitat—wild or semi-wild—are just as complex as nature, and just as wonderful. Their efforts come to fruition whenever we walk in nature, wherever life thrives, and whenever we, today, choose a similar gift to the future.

Eastern Tiger Swallowtail, Virginia. The special state license plate helped raise funds for habitat management.

Zebra Swallowtail, Shenandoah River State Park, Virginia. The grounds are managed with care to provide a mixture of groomed areas for family play alongside wild habitat. The burgundy blooming pawpaw tree is a host for caterpillars of the Zebra Swallowtail found nectaring on weedy flowers below.

Department of Conservation & Recreation
Virginia State Parks
Voted America's Best

Northern White-Skipper, Zion National Park, Utah. Plantings around the lodge are attractive to butterflies, which find their caterpillar host plants nearby in the wild areas.

Satyr Anglewing (ventral views) and Green Anglewing (dorsal views), Mount Hood National Forest, Oregon. A hiking bridge across the Hood River, at the Tamanawas Falls trailhead, took us to a small patch of streamside sunny boulders and wet sand that anglewings come to for brief visits.

Anise Swallowtail, Saddle Mountain State Park, Oregon. Anise Swallowtails are found at the very peak, as well as every clearing along the trail from the parking lot on up.

Moss's Elfin, Saddle Mountain State Park, Oregon. Moss's Elfin can often be found on the upper slopes of Saddle Mountain, in the vicinity of small sedums growing on rocks.

CASUAL HABITAT

An area behind the Douthat State Park headquarters buildings was lightly groomed, with many wild plants. Flowers were inconspicuous, but there were enough to slow down a butterfly, which led us on quite a chase.

Ebro, in the Florida panhandle, is just a crossroads, with a pleasant small rural motel. A vacant lot behind the next-door truck stop wasn't pretty, but still hosted a Palamedes Swallowtail. Here I first saw and photographed a Carolina Satyr.

Falcate Orangetip, female; Douthat State Park, Virginia.
Only the males have orange wing tips.

Carolina Satyr, Florida

Palamedes Swallowtail, Florida

Lindsey's Skipper, a vernal pool, wet in May but soon to dry up, in the Cascade-Siskiyou National Monument, Oregon. In dry areas, water is very attractive.

Mexican Yellow, Big Bend National Park, Texas. Noticing it nectaring in a prickly pear flower gave me a good laugh.

Vesta Crescent, Big Bend National Park, Texas. Tempted as we are to think of butterflies as pureness and delight, it's worthwhile to watch for them on poop and carrion, which they feed on.

LIKELY FLOWERS

Certain flowers seem especially likely to attract large numbers of butterflies, and quite a variety of species, becoming equally irresistible to the eager butterflier. A few of my favorites are on the next pages—any patch of these is worth checking. Your locale will have its own. Many other flowers will occasionally have nectarers, but are often quite bereft of such attention.

Rabbit brush, with Monarch, Oregon

Dogbane, with Hydaspe Fritillary, Ochoco National
Forest, Oregon

Clover with Atlantis Fritillaries, Réserve Faunique
La Vérendrye, Quebec

Cabbage White

Dun Skipper

Meadow Fritillary

Red-spotted Purple

Summer Azure

Aphrodite Fritillaries, Eastern Tiger Swallowtail

In one hour, one clump of Joe-Pye weed in Pennsylvania yielded the seven species above and one on the right.

Monarch on Joe-Pye weed, Allegheny National Forest, Pennsylvania

DOUBLETAIL

Two days of chasing butterflies through the hot, dry Siskiyous left me totally exhausted, and I worried, needlessly, it turned out, that a technical error had damaged much of the day's camera work. I parked at a promising butterfly site, but just sat in my truck feeling too discouraged to get out and look on the flowers.

Then the bright yellow of the largest wings in the Pacific Northwest swirled up my windshield and down past a side window. Freshly energized, I sprang from the truck. Before I could reach for my cameras, this bright beauty landed softly on my cheekbone. Here was a true dilemma: Dare I reach for my cameras, change lenses, and try to photograph myself adorned with *Papilio multicaudatus*? And probably scare off the butterfly? No, some of the great moments in life are made to experience directly, not by photographing, and not to jeopardize by grabbing for cameras. In a few minutes my friend— as I now dared call the magical being—flew off to a nearby thistle and posed with the sun coming through its wings, giving me all the shots I wanted.

Two-tailed Tiger Swallowtail and Woodland Skippers on thistle, Cascade-Siskiyou National Monument, Oregon

Mike Patterson doing a transect, counting Oregon Silverspots in Siuslaw National Forest, Oregon, for the federal Fish and Wildlife Service as part of an effort to maintain habitat for this endangered population. A transect is a formal count along an established route, conducted at regular intervals.

Oregon Silverspot Fritillary on goldenrod, Siuslaw National Forest, Oregon

WATER, MUD, SAND, AND SCAT

Any wet mud or sand is worth checking—moisture or puddle on a road or trail, or the edge of a stream. Butterflies need the water and sometimes mineral salts.

A challenge to our sense of butterflies as delicate and sweet? Scat and carrion are worth checking. The amino acids found there are valuable foods.

Painted Lady on dog poop, Wahkiakum County, Washington

California Tortoiseshell on streamside mud, Eagle Cap Wilderness, Wallowa-Whitman National Forest, Oregon

Anicia Checkerspot on sandy mud, Wenatchee National Forest, Washington

LOCAL KNOWLEDGE

Butterfly populations can be very spotty, absent here, but abundant around the corner. But around which corner?

For instance, Sue Anderson took me to a hundred-yard-long colony of Arctic Skippers, a species I had not yet photographed. By myself, I might well have found the promising habitat of the general area, yet could easily have taken a different branch of the trail. Nothing but Sue's localized experience would have guided me to that specific stretch. She also took me to the sites of the following two pages.

Mourning Cloak on
Sue Anderson, Oregon

Arctic Skipper, Metolius Preserve, Deschutes Land Trust Preserve, Oregon

Sparse flowers on dry ground attracted these Acmon or Lupine Blues,
Deschutes National Forest, Oregon

Bramble Green Hairstreak, Acmon Blue, Echo Azure, Juniper Hairstreak, and Brown Elfin; Deschutes National Forest, Oregon

More Butterflies

Great Arctic, Tub Springs State Wayside, Oregon

European Skipperling, Ontario

Western Meadow Fritillary, Deschutes National Forest, Oregon

Silver-bordered Fritillary, Ochoco National Forest, Oregon

MELISSA BLUES

A drive from Washington to southern California netted a fine visit with family—and California butterflies new to my camera. Alone and homeward bound, I butterflied through hills above the Feather River. In the town of Crescent Mills, I pulled onto dusty gravel where an industrial road crosses a railway. No flowers, not much to look at, but I needed a break from driving. Watching the tawny grass, I saw two tiny Melissa Blues mating. I was able to gently approach them with a macro lens for a very close photograph.

Melissa Blues, mating pair; California

Silvery Blue, Cascade-Siskiyou National
Monument, Oregon

Silver-spotted Skipper, Durham, North Carolina

Large Marble, Cascade-Siskiyou National Monument, Oregon

Cabbage White on oregano, Astoria, Oregon

Margined Whites, mating pair; Wahkiakum County, Washington

David Branch, Washington

Greenish Blue, Oregon

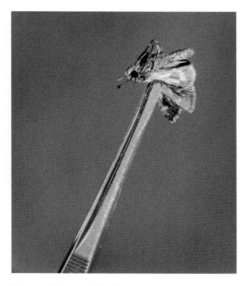

Sonora Skipper, Washington

Little Yellow on Fred Myers, Virginia

Anicia Checkerspot, Washington

For teaching on field trips, butterflies can be netted, viewed, and released unharmed. This allows people to see and discuss details that are hard to learn for the first time through binoculars. Then there's the smile factor. Butterflies often accept being placed on a person, and we all love it.

As one who works primarily photographically, I'm pleased with what I can accomplish and acutely aware of the limitations. A wild, free butterfly usually does not pose offering full view of each wing surface. And of course a photo does not provide tissue for DNA analysis or microscopic dissection. Specimen collections are crucial for most new knowledge, especially for taxonomic revisions. Some people worry that taking specimens will deplete a population, but thoughtful biologists and amateurs are not a hazard. Only the most irresponsible, unrestrained collectors do damage. We kill far more with habitat destruction, agriculture, and our fast cars than we can with nets.

Visual identification is an invaluable tool, critical for counting individuals and species. It's also fallible: several times I have thought I photographed a species I know well, only to find that it changed identity in my camera! I've heard similar stories from professional lepidopterists. I've also heard professionals make confident pronouncements that turned out to be wrong, according to my photographs.

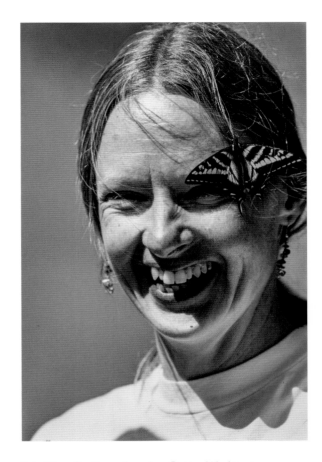

Pale Tiger Swallowtail on Ann Potter, Washington Department of Fish and Wildlife conservation biologist

Cabbage White on Andrew Emlen, environmental educator, with his son Connor Emlen-Petterson

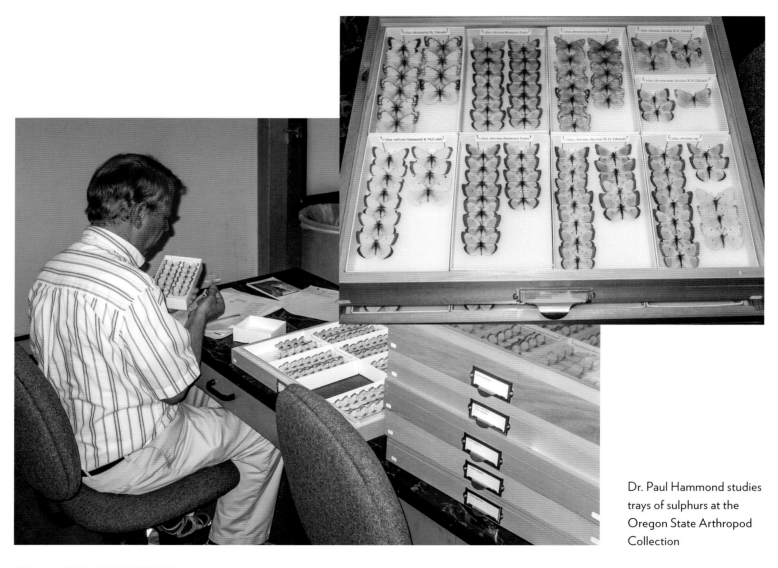

Dr. Paul Hammond studies
trays of sulphurs at the
Oregon State Arthropod
Collection

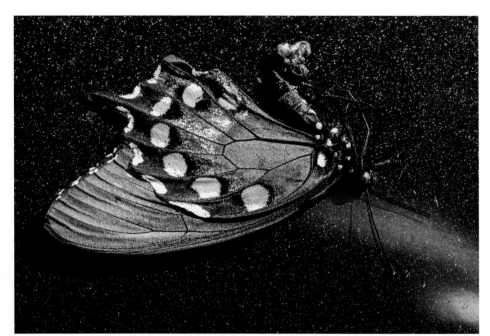

Pipevine Swallowtail, hit with my windshield; Texas

WILDNESS

Meadows, woods, and hills give me an experience of beauty and fascination, and I give to wildness my gratitude and thus my caring. When I am able to expand my consciousness sufficiently, my separation from my surroundings, from Nature, dissolves: instead of me perceiving it, it becomes my way of perceiving. And instead of being its observer, I become its consciousness, its awareness of itself. Now I am at home, in my ancestral way, my home of enchantment. Nature and I are within each other.

Hedgerow Hairstreak, Deschutes National Forest, Oregon

Empress Leilia on prickly pear, Saguaro National Park, Arizona

Indra Swallowtail, Deschutes River National Recreation Lands, Oregon

Gulf Fritillary, Spanish Harbor Wayside Park, Florida

Fritillary, likely Zerene, Oregon

Hydaspe Fritillary,
Washington

What's Next?

BUTTERFLY CONSERVATION

Usually a certain kind of butterfly depends on a few particular plants for its caterpillars to live on and eat. Without host plants there would be no caterpillars, then no butterflies. Many hosts are inconspicuous natives. As human activities expand across the landscape, such plants are finding fewer and fewer places to live. We use land for homes, for business and industry, for agriculture and commercial forestry. We bring in plants from other regions, sometimes because they are extra tough and vigorous—and then they spread widely, squeezing out native plants that butterflies depend on. Scotch broom, for instance, takes over whole hillsides, shading out small native herbs that are crucial for certain butterflies. Introduced grasses crowd out natives that many skipper butterfly caterpillars eat. Every time we divert water from a stream, every time we lower an aquifer by pumping, some of the previously moist low spots and stream banks dry up and no longer support the accustomed plants and caterpillars. So yes, there are fewer butterflies.

Cities displace habitat with buildings, pavement, and tightly controlled plantings. Economic activities displace habitat as well. As our population grows, we build more cities, have more economic activity, and displace more habitat. In the face of this we can each help. Find some spaces to encourage wild growth of native plants in your area. Part of a home yard, fencerows, roadsides, and any place you or your community choose to support habitat can be worthwhile. Certain conservation organizations buy larger tracts of habitat, and we can help fund them. Every such action contributes to maintaining the natural world and our opportunity to experience it.

San Jose, California, 2014

AGRICULTURE

Modern industrial agriculture all too often sacrifices every inch of habitat to crops and scorched earth. Heavy machinery, herbicides, and insecticides impoverish the habitat. Gentler farming, and a smaller population to be fed would be much better.

The corner of Paradise Avenue and Maze Boulevard? You can't make that up.

Near Tracy, California, 1999

CLIMATE CHANGE

Changing seasonal and weather patterns affect butterfly and moth populations. Shifts in the timing of precipitation and temperature affect which species can live where, and when. Some species are shifting northward or arriving earlier, making new appearance records. Others will find their survival here challenged by changes in temperature, vegetation, competitors, and predators. Abundant records of butterfly and moth sightings will help document these changes, and increase our understanding of it.

Mount Shasta's highest few thousand feet had year-round snow cover and glaciers until recently. In 1967 or '68, about the same late summer week, I climbed on snow from the dashed red line until stopped by an overhead snow cliff at the solid line.

Maintaining the full richness of butterfly species into the future will depend on maintaining their habitats—especially the wild, native host plants for caterpillars. Many of these are not often noticed or valued by people—they tend to be inconspicuous, or "brushy" or "weedy." Either we will learn to value them and provide a permanent place for them in the landscape, or future generations will see many fewer butterflies.

Mount Shasta, California, September 2014

IT'S ABOUT PEOPLE

People often remember noticing more butterflies when they were younger than they do now. Is this because as children our eyes were closer to the flowers, and we spent more time out of doors? Did we let our attention explore more? Children can still do that—and so can grown-ups. Dirty clothes and a few thorns? Sign me up!

Even if we, collectively, grievously wound Nature, in the long run Nature would nevertheless be fine. A few tens of millions of years and the complexities and subtleties would be restored. Nature has enough time. People don't. Our lives are now, and maybe in the next few thousand years, and whether or not those future lives are lived immersed in a soul-nurturing Nature, a Nature bigger than we are that keeps pulling on us and expanding us, that is a choice we make with our conduct today.

Yes, conserving wildness does temporarily help Nature, yet because of the sufficiency of Nature's healing time contrasted with the immediacy of our human experience, what is ultimately at stake is our human desire for lives as deep and rich as possible.

Have we finished with Nature? Are we through being fascinated, being challenged, and discovering? Are we through being swept away in awe, being comforted, and responding with art? I hope not. What's special about engaging wildness? It's greater than our knowledge. Wildness is even grander and stranger than our imaginations.

We were born into the opportunity to be aware, to care, and to create meaning in action. For the love of life, I say, play it for all it's worth. Is there any reason to settle for less?

Cindi with Eastern Tiger Swallowtail, Reedy River Falls Park, Greenville, South Carolina

Tyler Joki and Bob Pyle, butterflying class; Flying L Ranch, Washington. The boy's encounter with Bob sparked an enduring interest in nature. While serving as a Marine in Fallujah, nature helped him maintain perspective. Tyler went on to a graduate education in wildlife biology and now works as an environmental restoration project manager.

Frederick with Sachem Skipper, Virginia

Resources for Butterflying

BUTTERFLY AND GARDENING PRIMERS

• Robert Michael Pyle and Sarah Anne Hughes, *Handbook for Butterfly Watchers* (Houghton Mifflin, 1992). A page-turning journey into encountering butterflies.
• Scott Hoffman Black, Brianna Borders, Candace Fallon, Eric Lee-Mäder, and Matthew Shepherd, *Gardening for Butterflies: How You Can Attract and Protect Beautiful, Beneficial Insects* (Timber Press, 2016). Thorough information on garden design, procedures, plant descriptions, and even how to work within community laws and regulations.

FIELD GUIDES

Books that show many species on a page are valuable for looking through quickly to focus on what we're seeing, especially with range maps to help us ignore unlikelies. Guides covering half a continent or more are great for travel through unfamiliar regions. I keep many in my studio and take relevant ones into the field. Books function even in remote areas where there's no internet connection.

• James P. Brock and Kenn Kaufman, *Field Guide to Butterflies of North America* (Houghton Mifflin, 2003).
• Robert Michael Pyle, *Audubon Field Guide to North American Butterflies* (Knopf, 1981).

- Jeffrey Glassberg, *A Swift Guide to Butterflies of North America* (Princeton University Press, 2017).

- Peterson Field Guides:
  Western Butterflies, by Paul A. Opler and Amy Bartlett Wright, 1999.
  Eastern Butterflies, by Paul A. Opler and Vichai Malikul, 1992, 1998.

- Thomas J. Allen, James P. Brock, and Jeffrey Glassberg, *Caterpillars in the Field and Garden: A Field Guide to the Butterfly Caterpillars of North America* (Oxford University Press, 2005).

Regional guides are much more expansive about details in the areas they cover. Seek them out wherever you are. As a Pacific Northwesterner I rely heavily on these:

- Robert Michael Pyle and Caitlin C. LaBar, *Butterflies of the Pacific Northwest* (Timber Press, 2018). The new standard in my region, with updated taxonomy and range information. Thorough illustration and sensitive descriptions.
- Robert Michael Pyle, *Butterflies of Cascadia* (Seattle Audubon Society, 2002). Remains invaluable for its detailed descriptions, expansive discussions of taxonomic complexities and uncertainties, and details about the people and processes of discovery. An engaging read.
- David G. James and David Nunnallee, *Life Histories of Cascadia Butterflies* (Oregon State University Press, 2011). The discussions and illustrations of all of the butterflies of Washington state are informative and inspiring to butterfliers everywhere.
- Caitlin C. LaBar, *Butterflies of the Sinlahekin Wildlife Area: With Summaries of Other Okanogan County Butterflies* (Speyeria Press, 2017); www.northwestbutterflies.blogspot.com. Extraordinary illustrated detail of variation, locations, and host plants. Most of Washington state's species are covered, so it's widely useful throughout the Pacific Northwest.
- William Neill and Doug Hepburn, *Butterflies of the Pacific Northwest* (Mountain Press, 2007). Includes an effective introduction to butterfly biology.
- Crispin S. Guppy and Jon H. Shepard, *Butterflies of British Columbia* (University of British Columbia Press, 2001). British Columbia and adjacent provinces and states. A treasure of perspective and the most detail of any of the books listed herein, including chapters on conservation and biology, species accounts, and dot maps of butterfly records by subspecies.

You have bookshelves? There are many more books to love.

INTERNET RESOURCES

A book can only show at most a few individuals from a variable species. New research results show up slowly in books. When studying a photograph to identify a butterfly, after narrowing it down to a few species using books, I study online until my eyes fall out. A certain mark appears on a photographed wing. Is it present on most examples of that species? Is it absent or different on the other species to consider?

The Butterflies of America website (www.ButterfliesofAmerica.com) shows multiple specimens of nearly all North American subspecies, as well as many tropical and South American ones, and hosts the current version of Jonathan P. Pelham's *A Catalog of the Butterflies of the United States and Canada*.

BUTTERFLIERS

Experienced eyes helped me cut through a lot of confusion, to know what I'm seeing, to confirm or correct my attempts. Experienced field workers know where to go. How to find them?

- Many areas have local organizations worth connecting with for field trips and presentations. The one closest to me, for instance, is the vigorous Washington Butterfly Association, or WABA (www.wabutterflyassoc.org).
- The North American Butterfly Association (NABA) has many local chapters (www.naba.org).
- Audubon Society chapters often include butterfliers (www.audubon.org).
- Lepidopterists' Society, for studying moths and butterflies, provides excellent resources and members throughout the country (www.lepsoc.org).
- Government agencies and land trusts, schools, and colleges often have or know of biologists.

THE XERCES SOCIETY FOR INVERTEBRATE CONSERVATION

Reaching beyond conservation of popular mammals and birds, in its own words, "The society uses advocacy, education, and applied research to defend invertebrates." And they do it well (www.xerces.org).

FREE-FLYING BUTTERFLY GARDENS

Outdoor gardens planted with larval hosts and nectar sources concentrate butterflies from the surrounding areas. They give a very high reward for our visiting time. This book includes photos from butterfly gardens in a Florida state park, the National Butterfly Center in Texas, Arizona-Sonora Desert Museum, Franklin Canyon Park in California, and many home gardens.

BUTTERFLY HOUSES (CLOSED-IN ZOOS)

Charming, intimate experiences with butterflies provide a good start for close-up observation.

Making Butterfly Photographs

It's easy to love photographing butterflies, and you can have a go at it. It's fun, and also useful: With photos, we can study what we saw, seek advice on identification, and share with friends and community. Our sightings can contribute to the databases of location and timing—especially valuable in this time of rapid climate change. Our very best results contribute to publications and even become art on the wall.

In making the photographs for this book, I worked with two main situations. When butterflies were very active or just intolerant of my close approach, I worked at a distance with a long lens. And when they were engrossed in nectaring, mating, or sipping from wet earth or scat, they let me work within inches using a macro lens. Most of the photographs in this book were made with a Canon 300mm f/4 lens multiplied by a 1.4× tele-extender to a 420mm focal length, or by a Canon 100mm f/2.8 macro lens. Camera bodies started with the Canon A2 film body, replaced by a Digital Rebel XTi, and last and best, the Canon 5D Mark III.

In the field, long sleeves, long pants, and knee pads make working on the ground and plowing through shrubs and prickly grass more comfortable. Though I am stalking or chasing butterflies, intent on making my shot, I find it useful to have a gentle, nonpredatory mind-set. Maybe doing so just helps me move more slowly and smoothly, or maybe it somehow communicates that I am not a threat. Thoughts on clothing color are mixed. In his *Handbook for Butterfly Watchers*, Robert Michael Pyle notes that drab colors help avoid making butterflies fly away, yet sometimes butterflies are attracted to bright colors we wear. In my experience, style of motion is more important than color. I never know how close I can get, so I start photographing at a distance, step closer for more, and closer still, continuing until either I've done everything I desire, or the butterfly leaves.

Wild butterflies are seen in their own chosen surroundings, conducting their lives, uncaught and unmanipulated. Often I give a "butterfly's eye" view, seeing them eye-to-eye, in their own scale. Only natural light is used: I love light and shadow effects, and rim lighting and backlighting—especially "stained glass window" lighting with the sun coming right through the wings, even when dorsal and ventral markings are thus confounded or partly concealed.

What about flash photography? Flash records fine details of a butterfly even when posed at an inconvenient angle to the sun, or in overly dim light. However, comparison of the butterfly colors to surrounding plants, and the place of the butterfly in its habitat, are sacrificed. I leave such efforts to others.

TYPES OF PHOTOGRAPHS FOR EVIDENCE, FIELD GUIDES, AND ART

I use my camera to document what I see at a location, both butterflies and host plants. Most such photos are just evidence for identification, not good enough for publication or display. For field guide–type photos, I try for the upper (dorsal) surfaces spread out flat and entirely sharp, for the under (ventral) surfaces on one side, and for an angle giving much of both, to tie the two surfaces together on the same individual. Field-guide views are best made in lighting coming from somewhat behind the camera, to clearly show the wing markings. I always try for art photos, designing with the shapes of butterflies and their surroundings. Dramatic lighting coming through the wings gives a stained-glass-window effect.

CAMERAS AND SETTINGS

Many cameras, including even phones, have a chance of making photographs that are pleasing on a small screen and good for identifying the butterfly. To get the best results, though, requires the highest quality equipment selected for exactly these tasks. Several of Canon's competitors also make superb gear to consider. Certain features are especially valuable:

Image stabilizing allows us to hand-hold and follow the butterflies around, using moderate shutter speeds.

Camera reaction speed is critical. If it takes too long to open the shutter, the butterfly has often moved. Many cameras that make otherwise excellent images are too sluggish.

Boisduval's Blue, ventral, dorsal-ventral, and dorsal views; Deschutes River State Recreation Area, Oregon. This one made it easy, except that the dorsal view is only half flat.

Single-lens reflex (SLR) cameras with through-the-lens viewfinders remain unequalled for precise, instantaneous viewing of the focused butterfly and overall composition. Yet sometimes an electronic viewfinder or a screen on a mirrorless camera is a usable compromise.

Zoom lenses simplify and lighten your load, but their images are not quite the best.

A GPS location recorded in images' metadata makes the photos more easily useful for biologists. Be sure it will function outside the range of WiFi and cellular phone signals, and that your image-editing software can read the data.

Use a very small central focus point to choose what part of the butterfly to optimize and not have to fight with auto-focus catching nearby stems and leaves.

An aperture setting of f/11 or f/13 at an ISO of up to 800 gives me the most reliable results.

In both macro closeup and long telephoto work, focus is more sensitive than in ordinary photography: a tenth of an inch error ruins a butterfly photo. My imperfect eyes depend on auto-focus for quick work, and it too is imperfect, so I shoot three or four exposures of a pose, using a camera setting that refocuses for each shutter press. Each refocusing gives me a new chance at a superb result.

Manual focus is often best for extreme closeup work with a macro lens. Set the lens for the desired magnification (closeness), and move back and forth to get the right focus. This is much quicker and more effective than sending auto-focus hunting near and far.

Make good photographs of as many situations as you can, and be sure to look carefully at every one. Back in my studio reviewing fieldwork, I've found compositions enriched by spiders and flying insects that I hadn't noticed in the viewfinder. Sometimes when I've thought a butterfly had left a flower ahead of my shutter, I've found it shown in flight. My successes are usually from persistent, thoughtful work, or occasionally a quick, joyful catch.

Monarch leaving Joe-Pye weed. Five shots of nectaring were followed by this departure, taken at 1/320 second, f/6.3, ISO=400 with digital technology from 2006 or so.

Terms Commonly Used

WING SURFACES

- *Forewings* are the pair closer to the head.
- *Hindwings* are the pair closer to the tail.
- *Dorsal* surfaces are the upper side, best seen when the wings are held out flat.
- *Ventral* surfaces are the underneath side, best seen when the wings are folded.

LIFE STAGES

A female *oviposits*, or lays *eggs*, on or near *food plants* (*larval hosts*). These then hatch into *larvae* (*caterpillars*). When a caterpillar is finished growing, it forms a *chrysalis* (or *pupa*), in which it metamorphoses into an adult butterfly, which then *ecloses* (or emerges from the pupa).

SCIENTIFIC TERMINOLOGY

Lepidopterists are scientists who study butterflies and moths. The word is derived from "scale" and "wing" in ancient Greek. *Lepidopterology* is the field of study.

Latinized scientific names are governed by the International Code of Zoological Nomenclature (ICZN), so biologists around the world can communicate unambiguously about species and the more comprehensive, higher level groups into which they're organized. The types of names used in this book are:

- *Species*, a group of organisms that can breed with one another; the group is reproductively isolated from other species. Species names are printed in lowercase italics.
- *Subspecies* are distinctive subgroups of species. They're printed in lowercase italics.
- A *genus* is a group of closely related species, all descended from a common ancestor. Some scientists lump many species together into a single genus, whereas others split the same group of species into several genera, so you will see different arrangements in different publications. The genus name is written in italics and capitalized.

As an example, *Speyeria zerene hippolyta* means the genus *Speyeria*, the species *zerene*, and the subspecies *hippolyta*. If we can identify only the genus, but not the species, then we refer to it as *Speyeria* sp. (or *Speyeria* species).

- A *family* is one of the six major divisions of butterflies. Family names all end in *-idae*, and are capitalized but not italicized.
- A *subfamily* is division of a family. Subfamily names all end in *-inae*, and are capitalized, not italicized.

Species, Locations, and Dates

Scientific names follow *A Catalog of the Butterflies of the United States and Canada*, by Jonathan P. Pelham, online at www.ButterfliesofAmerica.com, version of July 1, 2017. Common names are not governed by any standard: the choices used in this book are largely informed by the Pacific Northwest lepidopterists among whom I have studied. Many workers will reasonably prefer alternatives. Further advancement of knowledge will continue to make revisions.

Many identifications in this book depended on additional views of the same individual or populations. Please contact the author with questions or concerns.

ii   Hoary (Zephyr) Anglewing, *Polygonia gracilis zephyrus*. Flat Rock Campground, Challis National Forest, near Sunbeam, Custer County, Idaho, July 31, 2014.

3   Western Tiger Swallowtail, *Papilio rutulus*. Myers home garden, Rosburg, Wahkiakum County, Washington, 1998.

4   California Tortoiseshell, *Nymphalis californica*. Near Pacific Crest Trail Baldy Creek Trailhead, Cascade-Siskiyou National Monument, Jackson County, Oregon, May 30, 2014.

5   Woodland Skipper, *Ochlodes sylvanoides*. Thea Pyle's home garden, Grays River, Wahkiakum County, Washington, August 4, 2013.

7   Mormon Metalmark, *Apodemia mormo*. Lower Klickitat River near Lyle, Fisher Hill Unit, Washington Department of Fish & Wildlife, Klickitat County, Washington, September 17, 2009.

9   Mourning Cloak, *Nymphalis antiopa*. Ontario State Park, Malheur County, Oregon, August 1998.

10   Fritillary, *Speyeria sp.* Steens Mountain Loop near Fish Camp, Harney County, Oregon, August 26, 2009.

11   Chalcedona Checkerspot, *Euphydryas chalcedona.* Northern Sonoma or Napa County, California, May 1997.

13   Silvery Blue, *Glaucopsyche lygdamus.* Morrison Creek Campground, Gifford Pinchot National Forest, Yakima County, Washington, August 5, 2012.

14   Anna's Blue, *Plebejus anna ricei.* Vicinity of Elk Lake north of Detroit, Willamette National Forest, Marion County, Oregon, July 23, 2009.

14   Echo Azure, *Celastrina echo,* with kayak guide Ginni Callahan. Price Island, Julia Butler Hansen National Wildlife Refuge for the Columbian White-tailed Deer, Wahkiakum County, Washington, Summer 1999.

15   Common Ringlet, *Coenonympha tullia.* Mack's Canyon vicinity, Deschutes River National Recreation Lands, Bureau of Land Management, Sherman County, Oregon, May 26, 2008.

16   Boisduval's Blue, *Icaricia icarioides.* Sawtooth Wilderness, Sawtooth National Forest, Custer County, Idaho, September 3, 2011.

17   Aphrodite Fritillary, *Speyeria aphrodite.* Twin Lakes Campground, Allegheny National Forest, Elk County, Pennsylvania, August 17, 2008.

21   Greenish Blue, *Icaricia saepiolus;* Western Branded Skipper, *Hesperia colorado idaho;* Mormon Fritillary, *Speyeria mormonia;* Field Crescent, *Phyciodes pulchella;* the same Crescent, second view; Sonora Skipper, *Polites sonora;* Dark Wood Nymph, *Cercyonis oetus;* Custer Motorway, Yankee Fork, Custer County, Idaho, August 1, 2014.

22   Northern Cloudywing, *Thorybes pylades.* Green Ridge near Camp Sherman, Jefferson County, Oregon, June 2, 2014.

22   Long-tailed Skipper, *Urbanus proteus.* Eno Commons–Indigo Creek Trail, Durham County, North Carolina, October 2, 2016.

22   Horace's Duskywing, *Erynnis horatius.* Huntley Meadows Park of Fairfax County, Virginia, September 1999.

22   Northern White-Skipper, *Heliopetes ericetorum.* Sun Lakes State Park, Grant County, Washington, September 22, 2013.

22   Common or White Checkered-Skipper, *Pyrgus communis* or *albescens.* NABA National Butterfly Center, Hidalgo County, Texas, April 23, 2014.

23 Propertius Duskywing, *Erynnis propertius*. O'Brien seep on Wimer Road, Josephine County, Oregon, June 11, 2005.

24 Common Checkered-Skipper, *Pyrgus communis*. Cottonwood Creek south of Fields Station, Harney County, Oregon, June 1, 2013.

25 Persius or Propertius Duskywing, *Erynnis persius* or *propertius*. Prairie Farm Meadow on Green Ridge near Camp Sherman, Jefferson County, Oregon, June 2, 2014.

26 Arctic Skipper, *Carterocephalus palaemon*. Metolius Preserve, Deschutes Land Trust Preserve near Camp Sherman, Jefferson County, Oregon, June 2, 2014.

26 Woodland Skipper, *Ochlodes sylvanoides*. Myers home garden, Astoria, Clatsop County, Oregon, September 11, 2008.

26 Juba Skipper, *Hesperia juba*. Near Bonanza, Challis National Forest, Custer County, Idaho, September 2, 2011.

26 Fiery Skipper, *Hylephila phyleus*. NABA National Butterfly Center, Hidalgo County, Texas, April 23, 2014.

26 Peck's Skipper, *Polites peckius*. Sam Owen Campground, Idaho Panhandle National Forests, Bonner County, Idaho, July 26, 2008.

26 Common Roadside-Skipper, *Amblyscirtes vialis*. Metolius Preserve, Deschutes Land Trust Preserve near Camp Sherman, Jefferson County, Oregon, June 2, 2014.

27 Rural Skipper, *Ochlodes agricola*. Wild Rogue Wilderness, Josephine County, Oregon, June 21, 2004.

28 Least Skipper, *Ancyloxypha numitor*. A home yard in Essex, Essex County, New York, August 9, 2008.

29 Woodland Skipper, *Ochlodes sylvanoides*. Myers home garden, Astoria, Clatsop County, Oregon, August 20, 2002.

30 Mountain Parnassian, *Parnassius smintheus*. Trail 640 from Stanley Lake, Custer County, Idaho, August 3, 2014.

30 Clodius Parnassian, *Parnassius clodius claudianus*. Probable location is Saddle Mountain State Park, Clatsop County, Oregon, about July 2007.

30 Anise Swallowtail, *Papilio zelicaon*. Saddle Mountain State Park, Clatsop County, Oregon, June 9, 2014.

30 Indra Swallowtail, *Papilio indra*. Boat ramp, Mack's Canyon, Deschutes River Recreation Area, Bureau of Land Management, Sherman County, Oregon, April 25, 2013.

31   Western Tiger Swallowtail, *Papilio rutulus*. Cape Disappointment State Park, Pacific County, Washington, May 21, 2000.

32   Anise Swallowtail, *Papilio zelicaon*. R. M. Pyle's home garden, Grays River, Wahkiakum County, Washington, August 17, 2014.

33   Western Tiger Swallowtail, *Papilio rutulus*. Headquarter grounds, Malheur National Wildlife Refuge, Harney County, Oregon, June 3, 2013.

34   Eastern Tiger Swallowtail, *Papilio glaucus,* yellow female. Huntley Meadows Park of Fairfax County, Virginia, September 1999.

34   Eastern Tiger Swallowtail, *Papilio glaucus,* black female. Home yard in Falls Church, Fairfax County, Virginia, August 27, 2010.

34   Palamedes Swallowtail, *Papilio palamedes*. Suwanee Canal area headquarters, Okefenokee National Wildlife Refuge, Charlton County, Georgia, April 6, 2014.

35   Eastern Tiger Swallowtail, *Papilio glaucus,* yellow female. Potomac Falls, Scott's Run Nature Preserve of Fairfax County, Virginia, September 14, 1999.

36   Eastern Giant Swallowtail, *Heraclides cresphontes*. Caley Reservation, Lorain County Metro Parks, Ohio, August 18, 2008.

37   Black Swallowtail, *Papilio polyxenes*. NABA National Butterfly Center, Hidalgo County, Texas, April 23, 2014.

38   Clouded Sulphur and Orange Sulphur, *Colias philodice* and *C. eurytheme*. Farm near Hop Bottom near Nicholson, Susquehanna County, Pennsylvania, August 13, 2008.

38   Clouded Sulphur, *Colias philodice,* white form. Thunder Basin National Grassland, Weston or Campbell County, Wyoming, August 24, 2008.

38   Western Sulphur, *Colias occidentalis*. Metolius Preserve, Deschutes Land Trust Preserve near Camp Sherman, Jefferson County, Oregon, June 2, 2015.

38   Cloudless Sulphur, *Phoebis sennae*. Mount Lemon, Coronado National Forest, Pima County, Arizona, August 4, 2000.

39   Pink-edged Sulphur, *Colias interior*. Réserve Faunique La Vérendrye, Quebec. August 4, 2008.

40   Orange Sulphur, *Colias eurytheme*. Motel grounds in Interior, Jackson County, South Dakota, August 21, 2008.

41   Dainty Sulphur, *Nathalis iole*. NABA National Butterfly Center, Hidalgo County, Texas, April 23, 2014.

42   Cabbage White, *Pieris rapae*. Myers home garden, Astoria, Clatsop County, Oregon, September 11, 2012.

42   Large Marble, *Euchloe ausonides*. Near Page Springs camp, Steens Mountain Recreation Lands, Bureau of Land Management, Harney County, Oregon, June 2, 2013.

42   Great Southern White, *Ascia monuste*. Merritt Island National Wildlife Refuge, Brevard County, Florida, April 7, 2014.

42   Sara's Orangetip, *Anthocharis sara*. Green Ridge near Camp Sherman, Jefferson County, Oregon, June 2, 2014.

42   Pine White, *Neophasia menapia*. Beauty Creek campground, Coeur d'Alène National Forest, Kootenai County, Idaho, August 28, 2011.

42   Western White, *Pontia occidentalis*. Trail 640 from Stanley Lake, Custer County, Idaho, August 3, 2014.

43   Becker's White, *Pontia beckerii*. Price Canyon Recreation Area, Bureau of Land Mangement, Carbon County, Utah, August 8, 2012.

44   Great Southern White, *Ascia monuste*. Spanish Harbor Wayside Park, Florida Department of Transportation, Monroe County, Florida, April 8, 2014.

45   Arctic White, *Pieris angelika*. Campground of Porcupine Creek State Recreation Site, Valdez Cordova Census Area, Alaska, July 5, 2010.

46   Purplish Copper, *Lycaena helloides*. Clatsop Plains, Clatsop County, Oregon, August 29, 2002.

46   Lustrous Copper, *Lycaena cupreus*. Gaylor Lake trail north from Tioga Pass, Yosemite National Park, Tuolomne County, California, late June 2007.

46   Harvester, *Feniseca tarquinius*. Sky Meadows, Shenandoah Mountains, Sky Meadows State Park, Fauquier County, Virginia, September 1, 2008. Photograph by Frederick L. Myers.

46   Blue Copper, *Lycaena heteronea*. Vicinity of Bonanza, Challis National Forest, Custer County, Idaho, September 2, 2011.

46   Edith's Copper, *Lycaena editha*. Walton Lake vicinity, Ochoco National Forest, Crook County, Oregon, July 18, 2013.

46   American Copper, *Lycaena phlaeas*. Big Meadows, Shenandoah National Park, Madison County, Virginia, June 18, 1999.

47   Purplish Copper, *Lycaena helloides*. Mack's Canyon vicinity, Deschutes River National Recreation Lands, Bureau of Land Management, Sherman County, Oregon, May 25, 2008.

47   Tailed Copper with Fred Myers, *Lycaena arota*. Vicinity of Yellowpine campground, Ashley National Forest, Summit County, Utah, August 17, 2012.

48   Brown Elfin, *Callophrys augustinus*. City yard, Astoria, Clatsop County, Oregon, May 28, 2004.

48   Dusky-blue Groundstreak, *Calycopis isobeon*. NABA National Butterfly Center, Hidalgo County, Texas, April 23, 2014.

48   Western Pine Elfin, *Callophrys eryphon*. Black Butte summit trail, Jefferson County, Oregon, June 1, 2014.

48   Coral Hairstreak, *Satyrium titus*. Near Bonanza, Challis National Forest, Custer County, Idaho, September 2, 2011.

48   White-M Hairstreak, *Parrhasius m-album*. Big Meadows, Shenandoah National Park, Madison County, Virginia, June 18, 1999.

48   Eastern Tailed-Blue, *Cupido comyntas*. Herrick Lake Forest Preserve, Forest Preserve District of Dupage County, Illinois, August 19, 2008.

49   Juniper Hairstreak, *Callophrys gryneus*. Near Page Springs camp, Steens Mountain Recreation Lands, Bureau of Land Management, Harney County, Oregon, June 4, 2013.

50   Gray Hairstreak, *Strymon melinus*. Metolius Preserve, Deschutes Land Trust Preserve near Camp Sherman, Jefferson County, Oregon, June 3, 2014.

51   Golden Hairstreak, *Habrodais grunus*. Rogue River Gorge vicinity, Rogue River National Forest, Jackson County, Oregon, early August 2003.

52   Echo Azures, *Celastrina echo;* Western Tailed-Blue, *Cupido amyntula*. Green Ridge near Camp Sherman, Jefferson County, Oregon, June 2, 2014.

52   Echo Azure, *Celastrina echo*. Green Ridge near Camp Sherman, Jefferson County, Oregon, June 2, 2014.

52   Melissa Blue, *Plebejus melissa*. Rapid River trail south of Riggins, Nez Perce National Forest, Idaho County, Idaho, September 8, 2011.

52   Marine Blue, *Leptotes marina*. Mount Lemon, Coronado National Forest, Pima County, Arizona, August 4, 2000.

52   Silvery Blue, *Glaucopsyche lygdamus*. Mack's Canyon vicinity, Deschutes River National Recreation Lands, Bureau of Land Management, Sherman County, Oregon, April 24, 2013.

53   Melissa Blue, *Plebejus melissa*. Black Canyon of the Gunnison National Park, Montrose County, Colorado, August 10, 2012.

54   Unidentified Blue. Muncho Lake Provincial Park, British Columbia, July 20, 2010.

55   Boisduval's Blue, *Icaricia icarioides*. Flat Rock Campground, Challis National Forest, near Sunbeam, Custer County, Idaho, September 3, 2011.

56   Greenish Blues, *Icaricia saepiolus*. Cow Meadow Camp, Cascade Lakes Highway, Deschutes National Forest, Deschutes County, Oregon, July 15, 2012.

57   Northern Blue, *Plebejus idas*. Congdon Government Campground, Yukon, July 18, 2010.

58   Unidentified Calephelis, *Calephelis sp.* Coatzacoalcos, Estado de Veracruz, March 1999.

58   Fatal Calephelis, *Calephelis nemesis*. Franklin Canyon Park, Santa Monica Mountains Conservancy, Los Angeles County, California, October 24, 1995.

58   Fatal Calephelis, *Calephelis nemesis*. NABA National Butterfly Center, Hidalgo County, Texas, April 23, 2014.

59   Mormon Metalmark, *Apodemia mormo*. Fisher Hill Unit, Washington Department of Fish & Wildlife, Klickitat County, Washington, September 17, 2009.

60   American Snout, *Libytheana carinenta*. Coronado National Forest, Cochise County, Arizona, August 5, 2000.

60   American Snout, *Libytheana carinenta*. Few's Ford, Eno River State Park, Orange County, North Carolina, May 26, 2014. Photograph by Frederick L. Myers.

61   American Snout, *Libytheana carinenta*. Few's Ford, Eno River State Park, Orange County, North Carolina, May 26, 2014. Photograph by Frederick L. Myers.

62   Queens, *Danaus gilippus*. NABA National Butterfly Center, Hidalgo County, Texas, April 23, 2014. Two photos, dorsal and ventral.

62   Monarch, *Danaus plexippus*. Santuario de la Mariposa Monarca el Rosario near Angangueo, Michoacán, March 1999.

62   Monarch, *Danaus plexippus*. Big Meadows, Shenandoah National Park, Madison
        County, Virginia, June 18, 1999.

63   Queen, *Danaus gilippus*. NABA National Butterfly Center, Hidalgo County, Texas,
        April 23, 2014.

64   Gulf Fritillary, *Agraulis vanillae*. Spanish Harbor Wayside Park, Florida Department of
        Transportation, Monroe County, Florida, April 8, 2014.

64   Zebra Longwing, *Heliconius charithonia*. Bahia Honda State Park, Monroe County,
        Florida, April 10, 2014.

64   Variegated Fritillary, *Euptoieta claudia*. Rabbit Mountain, Boulder County Parks,
        Colorado, August 12, 2012.

64   Fritillary, likely Zerene, *Speyeria zerene*. Big Summit Prairie vicinity, Ochoco National
        Forest, Crook County, Oregon, July 16, 2013.

64   Hydaspe Fritillary, *Speyeria hydaspe*. Road west from Big Summit Prairie, Ochoco
        National Forest, Crook County, Oregon, July 16, 2013.

64   Meadow Fritillary, *Boloria bellona*. A home yard in Essex, Essex County, New York,
        August 9, 2008.

64   Western Meadow Fritillary, *Boloria epithore*. Wonderland Trail near Maple Creek trail-
        head, Mount Rainier National Park, Lewis County, Washington, July 7, 2014.

65   Zebra Longwing, *Heliconius charithonia*. Bahia Honda State Park, Monroe County,
        Florida, April 10, 2014.

66   Variegated Fritillary, *Euptoieta claudia*. Chisos Basin, Big Bend National Park, Brewster
        County, Texas, April 26, 2014.

67   Great Spangled Fritillary, *Speyeria cybele*. Woods NW of Big Summit Prairie, Ochoco
        National Forest, Crook County, Oregon, July 18, 2013.

68   Lorquin's Admiral, *Limenitis lorquini*. Vicinity entrance to Glacier Lodge, Inyo National
        Forest near Big Pine, Inyo County, California, July 12, 2017.

68   White Admiral, *Limenitis arthemis*. Hidden Lake Trail near Northway, Alaska, July 5,
        2010.

68   California Sister, *Adelpha californica*. Vicinity of Paxton, Plumas National Forest, Plumas
        County, California, August 28, 1999.

68   Hackberry Emperor, *Asterocampa celtis*. Roadside, Morrison, Whiteside County, Illinois,
        August 20, 2008.

69   California Sister, *Adelpha californica*. Rogue River trail west of Grave Creek, Siskiyou National Forest, Josephine County, Oregon, June 21, 2004.

70   Tawny Emperor, *Asterocampa clyton*. NABA National Butterfly Center, Hidalgo County, Texas, April 23, 2014.

71   Lorquin's Admiral, *Limenitis lorquini*. Ecola Creek Forest Reserve, City of Cannon Beach, Clatsop County, Oregon, August 8, 2013.

72   West Coast Lady, *Vanessa annabella*. Bigelow Lakes Basin, Oregon Caves National Mounument and Preserve, Josephine County, Oregon, July 1, 2017.

72   American Lady, *Vanessa virginiensis*. Bigelow Lakes Basin, Oregon Caves National Mounument and Preserve, Josephine County, Oregon, July 1, 2017.

72   Red Admirable, *Vanessa atalanta*. Soda Mountain, Cascade-Siskiyou National Monument, Jackson County, Oregon, May 28, 2006.

73   Red Admirable, *Vanessa atalanta*. Wallowa Lake State Park, Wallowa County, Oregon, August 6, 2014.

74–76   Painted Ladies, *Vanessa cardui*, on asters. Myers home yard, Wahkiakum County, Washington, Summer 1995.

77   Mourning Cloak, *Nymphalis antiopa*. Mack's Canyon vicinity, Deschutes River National Recreation Lands, Bureau of Land Management, Sherman County, Oregon, May 26, 2008.

77   California Tortoiseshell, *Nymphalis californica*. Gifford Pinchot National Forest, southwest of Mount Adams, Skamania County, Washington, August 8, 2001.

77   Milbert's Tortoiseshell, *Agalais milberti*. Heaven's Gate, Hells Canyon National Recreation Area, Idaho County, Idaho, September 9, 2011. Two photographs.

78   Milbert's Tortoiseshell, *Agalais milberti*. Chief Joseph Pass, Bitterroot National Forest, Ravalli County, Montana, August 31, 2011.

79   Question Mark, *Polygonia interrogationis*. NABA National Butterfly Center, Hidalgo County, Texas, April 23, 2014.

79   Comma Anglewing, *Polygonia comma*. Catawba Falls Trail, Pisgah National Forest, McDowell County, North Carolina, April 3, 2014.

79   Green Anglewing, *Polygonia faunus*. Bigelow Lakes Basin, Oregon Caves National Monument and Preserve, Josephine County, Oregon, July 17, 2017.

80 Hoary (Zephyr) Anglewing, *Polygonia gracilis zephyrus*. Wizard Falls Hatchery, Department of Fish & Wildlife, Jefferson County, Oregon, September 21, 2014.

81 Common Buckeye, *Junonia coenia*. Merritt Island National Wildlife Refuge, Brevard County, Florida, April 7, 2014. Two views.

81 Common Buckeye, *Junonia coenia*. Alum Rock Park, City of San Jose, Santa Clara County, California, August 29, 1995.

81 Mangrove Buckeye, *Junonia genoveva*. Ding Darling National Wildlife Refuge, Sanibel Island, Lee County, Florida, April 12, 2014.

82 Mangrove Buckeye, *Junonia genoveva*. Ding Darling National Wildlife Refuge, Sanibel Island, Lee County, Florida, April 12, 2014.

83 Edith's Checkerspot, *Euphydryas editha*. Cow Meadow Camp, Cascade Lakes Highway, Deschutes National Forest, Deschutes County, Oregon, July 13, 2012.

83 Snowberry Checkerspot, *Euphydryas colon*. Porter Creek north from Deep Creek campground, east of Big Summit Prairie, Ochoco National Forest, Oregon, July 17, 2013.

83 Sagebrush Checkerspot, *Chlosyne acastus*. Mack's Canyon vicinity, Deschutes River National Recreation Lands, Bureau of Land Management, Sherman County, Oregon, May 21, 2006.

83 Leanira Checkerspot, *Chlosyne leanira*. Rogue River trail west of Grave Creek, Siskiyou National Forest, Josephine County, Oregon, June 21, 2004.

83 Bordered Patch, *Chlosyne lacinia*. NABA National Butterfly Center, Hidalgo County, Texas, April 23, 2014.

84 Anicia Checkerspot, *Euphydryas colon,* Thea Linnea Pyle. Swakane Canyon, Wenatchee National Forest, Wenatchee County, Washington, June 27±, 1999.

85–86 Chalcedona Checkerspot, *Euphydryas chalcedona*. Near Pacific Crest Trail Baldy Creek Trailhead, Cascade-Siskiyou National Monument, Jackson County, Oregon, May 30, 2014.

87 Mylitta Crescent, *Phyciodes mylitta,* female. Heaven's Gate, Hells Canyon National Recreation Area, Idaho County, Idaho, September 9, 2011.

87 Mylitta Crescent, *Phyciodes mylitta,* male. Vicinity of Allingham campground, Camp Sherman, Deschutes National Forest, Jefferson County, Oregon, June 27, 2013.

87 Field Crescent, *Phyciodes pulchella*. Vicinity of Elk Lake north of Detroit, Willamette National Forest, Marion County, Oregon, July 23, 2009.

87 Pale Crescent, *Phyciodes pallida*. Mack's Canyon vicinity, Deschutes River National Recreation Lands, Bureau of Land Management, Sherman County, Oregon, May 18, 2012.

87 Texan Crescent, *Anthanassa texana*. NABA National Butterfly Center, Hidalgo County, Texas, April 23, 2014.

88 Mylitta Crescent, *Phyciodes mylitta*. Sunshine Lake, Eagle Cap Wilderness, Wallowa County, Oregon, September 12, 2007.

89 Common Ringlet, *Coenonympha tullia*. A home yard in Essex, Essex County, New York, August 10, 2008.

89 Common Alpine, *Erebia epipsodea*. Two to three kilometers west of Summit Lake town, British Columbia, July 21, 2010.

89 Dark Wood Nymph, *Cercyonis oetus*. Aspenglen Campground, Rocky Mountain National Park, Larimer County, Colorado, August 12, 2012.

89 Northern Pearly-eye, *Lethe anthedon*. A home yard in Essex, Essex County, New York, August 9, 2008.

89 Chryxus Arctic, *Oeneis chryxus*. Trail 640 from Stanley Lake, Custer County, Idaho, August 3, 2014.

89 Great Arctic, *Oeneis nevadensis*. Vicinity of Trillium Lake, Mount Hood National Forest, Clackamas County, Oregon, July 18, 2008.

90 Great Arctic, *Oeneis nevadensis*. Cascade Lakes Highway, Deschutes National Forest, Deschutes County, Oregon, July 15, 2012.

91 Common Ringlet, *Coenonympha tullia*. A home yard in Essex, Essex County, New York, August 9, 2008.

92 Common Wood Nymph, *Cercyonis pegala*. Beauty Creek campground, Coeur d'Alène National Forest, Kootenai County, Idaho, August 29, 2011.

94 Acmon/Lupine Blue, *Icaricia acmon/lupini*. Green Ridge near Camp Sherman, Jefferson County, Oregon, September 20, 2014.

94 Oregon Swallowtail, *Papilio machaon oregonia*. Mack's Canyon vicinity, Deschutes River National Recreation Lands, Bureau of Land Management, Sherman County, Oregon, April 25, 2013. Two views.

95 Western Tiger Swallowtail, *Papilio rutulus*. Wahkiakum County, Washington, June 1995.

95 Anise Swallowtail, *Papilio zelicaon*. Wahkiakum County, Washington, July 1995.

95   Western Meadow Fritillary, *Boloria epithore*. Dorsal view from Kings Canyon National Park, California, June 13, 2002. Ventral view from a specimen, Wahkiakum County, Washington, collected in the late 1980s.

95   Painted Lady, *Vanessa cardui*. Myers home yard, Wahkiakum County, Washington, Summer 1995.

95   West Coast Lady, *Vanessa annabella*. Cascade Head Nature Conservancy Preserve, Tillamook County, Oregon, August 9, 1995.

96   Northern Blue, *Plebejus idas*. A few miles westbound from Teton Pass, Teton County, Wyoming, August 25, 2008.

96   Western Pygmy Blue, *Brephidium exilis*. Krumbo Reservoir Road, Malheur National Wildlife Refuge, Harney County, Oregon, September 15, 2016.

96   Nabokov's Satyr, *Cyllopsis pyracmon*. Vicinity of Ramsay Canyon, Coronado National Forest, Cochise County, Arizona, about August 5, 2000.

96   Coronis Fritillary, *Speyeria coronis*. Dismal Swamp, Modoc National Forest, California, June 23, 2018.

97   Cassius Blue, *Leptotes cassius*. Bahia Honda State Park, Monroe County, Florida, April 10, 2014.

98   Monarch, *Danaus plexippus*. Green Ridge near Camp Sherman, Jefferson County, Oregon, September 21, 2014.

100   Monarch caterpillar, *Danaus plexippus*. Rogue River trail at Foster Bar, near Agnes, Rogue River National Forest, Curry County, Oregon, June 25, 2004.

100   Monarch chrysalis, *Danaus plexippus*. Raised by David James; photograph by David G. James.

100   Owl Butterfly, *Caligo species*. Mariposarium in Mindo, Ecuador.

100   Monarch, *Danaus plexippus*. Mariposarium in Mindo, Ecuador.

101   Monarch, *Danaus plexippus*. Private home, Falls Church, Fairfax County, Virginia, August 28, 2010.

103   Santuario de la Mariposa Monarca el Rosario, near Angangueo, Michoacán, March 1999.

105   Monarchs, *Danaus plexippus*. Santuario Mariposa Monarca Sierra Chincua, near Angangueo, Michoacán, March 1999.

106   Monarchs, *Danaus plexippus*. Santuario Mariposa Monarca Sierra Chincua, near Angangueo, Michoacán, March 1999.

107   Monarchs, *Danaus plexippus*. Santuario Mariposa Monarca Sierra Chincua, near Angangueo, Michoacán, March 1999.

108   Monarchs, *Danaus plexippus*. Santuario Mariposa Monarca Sierra Chincua, near Angangueo, Michoacán, March 1999.

109   Monarchs, *Danaus plexippus*. Santuario de la Mariposa Monarca el Rosario near Angangueo, Michoacán, March 1999.

109   Monarchs, *Danaus plexippus,* with Robert Michael Pyle. Santuario de la Mariposa Monarca el Rosario, near Angangueo, Michoacán, March 1999.

110   Monarchs, *Danaus plexippus*. Santuario de la Mariposa Monarca el Rosario, near Angangueo, Michoacán, March 1999.

111   Monarchs, *Danaus plexippus*. Santuario Mariposa Monarca Sierra Chincua, near Angangueo, Michoacán, March 1999.

113   Monarchs mating, *Danaus plexippus*. Santuario Mariposa Monarca Sierra Chincua, near Angangueo, Michoacán, March 1999.

113   Monarchs, *Danaus plexippus,* and schoolchildren. Santuario Mariposa Monarca Sierra Chincua, near Angangueo, Michoacán, March 1999.

114   Monarch, *Danaus plexippus*. Key West Art and Historical Society's Custom House Museum, Monroe County, Florida, April 9, 2014.

115–116 Monarchs, *Danaus plexippus*. Sperling Preserve, Goleta, Santa Barbara County, California, October 19, 1995.

117   Monarch wings, *Danaus plexippus*. Sperling Preserve, Goleta, Santa Barbara County, California, October 19, 1995.

119   Tawny Emperors, *Asterocampa clyton*. NABA National Butterfly Center, Hidalgo County, Texas, April 23, 2014.

120   Tawny Emperor, *Asterocampa clyton*. Identification based on having seen the butterfly in the group on the previous page before it was caught. NABA National Butterfly Center, Hidalgo County, Texas, April 23, 2014.

121   Common Buckeye, *Junonia coenia*. Eno Commons-Indigo Creek Trail, Durham County, North Carolina, October 1, 2016.

121 Eastern Tiger Swallowtail, *Papilio glaucus*. Eno River State Park, Orange or Durham County, North Carolina, April 16, 2016.

122 Arctic Fritillary, *Boloria chariclea*. Shorthorn Trail on Mount Adams, from Morrison Creek Campground, Gifford Pinchot National Forest, Yakima County, Washington, August 5, 2012.

123 Red-spotted Purple form of White Admiral, *Limenitis arthemis astyanax*. Farm yard in Hop Bottom, Susquehanna County, Pennsylvania, August 13, 2008.

124 Pale Tiger Swallowtail, *Papilio eurymedon*. Roadside at entrance to Deep Creek campground, east of Big Summit Prairie, Ochoco National Forest, Oregon, July 17, 2013.

125 Spider with Sulphur, *Colias sp*. Roadside near Toppenish National Wildlife Refuge on U.S. route 97, Yakima County, Washington, October 9, 1995.

126 Spider and likely a Western Branded Skipper, *Hesperia colorado*. Northern edge of Big Summit Prairie, Ochoco National Forest, Crook County, Washington, September 12, 1996.

127 Woodland Skipper and Spider, *Ochlodes sylvanoides*. Cascade-Siskiyou National Monument, Jackson County, Oregon, August 9–10, 2003.

128 Yehl Skipper, *Poanes yehl*. Point Lookout State Park, Saint Mary's County, Maryland, September 13, 1999.

129 Common Ringlet, *Coenonympha tullia*. Mack's Canyon, Deschutes River Recreation Area, Bureau of Land Management, Sherman County, Oregon, May 25, 2008.

130 California Tortoiseshell with spider, *Nymphalis californica*. Cascade-Siskiyou National Monument, Jackson County, Oregon, May 30, 2014.

131 Unidentifiable Fritillary, *Speyeria sp*. Near Bonanza, Challis National Forest, Custer County, Idaho, September 2, 2011.

132 Western Branded Skipper, *Hesperia colorado*, female, with ambush bug. Continental Lake, Bureau of Land Management, Winnemucca District, Humboldt County, Nevada, August 11±, 2000.

133 Milbert's Tortoiseshell, *Aglais milberti*. Center Patrol Road, Malheur National Wildlife Refuge, Harney County, Oregon, June 3, 2013.

134 Hoary (Zephyr) Anglewing, *Polygonia gracilis zephyrus*. Near Bonanza, Challis National Forest, Custer County, Idaho, September 2, 2011.

135 Hoary (Zephyr) Anglewing, *Polygonia gracilis zephyrus*. Santiam Pass, Deschutes National Forest, Jefferson County, Oregon, July 24, 2009.

136 Ornythion Swallowtail caterpillar, *Papilio ornythion*. NABA National Butterfly Center, Hidalgo County, Texas, April 23, 2014.

136 Monarch, *Danaus plexippus*. Big Meadows, Shenandoah National Park, Madison County, Virginia, June 18, 1999.

137 Viceroy, *Limenitis archippus*. Caley Reservation, Lorain County Metro Parks, Ohio, August 18, 2008.

138–139 Painted Lady, *Vanessa cardui*. Myers home yard, Wahkiakum County, Washington, Summer 1995.

140 Pine White, *Neophasia menapia*. Indian Trees campground, Bitterroot National Forest, Ravalli County, Montana, August 31, 2011.

141 Woodland Skipper, *Ochlodes sylvanoides*. Fisher Hill Unit, Washington Department of Fish & Wildlife, Klickitat County, Washington, September 17, 2009.

142 Pale Tiger Swallowtail, *Papilio eurymedon*. Alum Rock Park, City of San Jose, Santa Clara County, California, August 28, 1995.

143 Anise Swallowtail, *Papilio zelicaon*. Myers home yard, Wahkiakum County, Washington, 1996.

144 Anise Swallowtail, *Papilio zelicaon*. Alcyon Farm, Middle Valley, Skamokawa, Wahkiakum County, Washington, July 6, 2005.

145 Queens, *Danaus gilippus*. Arizona-Sonora Desert Museum outdoor butterfly gardens, Tucson, Pima County, Arizona, August 2, 2000.

146 Gulf Fritillaries, *Agraulis vanillae*. Del Rio Beach, Santa Cruz County, California, August 30, 2014.

147 Cabbage White, *Pieris rapae*. Myers home garden, Astoria, Clatsop County, Oregon, September 11, 2012, August 2, 2000.

148 Cabbage White, *Pieris rapae*. Myers home garden, Astoria, Clatsop County, Oregon, July 2, 2011.

149 Cabbage White, *Pieris rapae*. Myers home garden, Astoria, Clatsop County, Oregon, May 26, 2012.

150 Great Basin Fritillaries, *Speyeria egleis*. Mount Pisgah vicinity lookout, Ochoco National Forest, Wheeler County, Oregon, July 19, 2003.

151   Spicebush Swallowtail, *Papilio troilus*. Huntley Meadows Park of Fairfax County, Virginia, September 1999.

151   Spicebush Swallowtail caterpillar, *Papilio troilus*. Shenandoah River State Park, Warren County, Virginia, September 19, 1999.

152   Hydaspe Fritillary, *Speyeria hydaspe*. Big Summit Prairie vicinity, Ochoco National Forest, Crook County, Oregon, July 18, 2013.

154   Juba Skipper, *Hesperia juba*. Mack's Canyon vicinity, Deschutes River National Recreation Lands, Bureau of Land Management, Sherman County, Oregon, April 22, 2013.

155   Gray Hairstreak, *Strymon melinus*. Mack's Canyon vicinity, Deschutes River National Recreation Lands, Bureau of Land Management, Sherman County, Oregon, April 23, 2013.

156   Anise Swallowtail, *Papilio zelicaon;* Oregon Swallowtail, *Papilio machaon oregonia*. Mack's Canyon boat ramp, Deschutes River National Recreation Lands, Bureau of Land Management, Sherman County, Oregon, April 25, 2013.

158   Eastern Tiger Swallowtail, *Papilio glaucus*. Home yard in Falls Church, Fairfax County, Virginia, August 24, 2010.

159   Eastern Tiger Swallowtail, *Papilio glaucus*. Home yard, Falls Church, Fairfax County, Virginia, August 24, 2010.

160   Zebra Swallowtail, *Eurytides marcellus*. Shenandoah River State Park, Warren County, Virginia, May 12, 2013.

161   Northern White-Skipper, *Heliopetes ericetorum*. Zion Lodge, Zion National Park, Washington County, Utah, August 27, 2012.

162   Ventral views of Satyr Anglewing, *Polygonia satyrus;* dorsal views of Green Anglewing, *Polygonia faunus*. Tamanawas Falls Trail crossing the Hood River, Mount Hood National Forest, Hood River County, Oregon, September 19, 2014.

163   Anise Swallowtail, *Papilio zelicaon*. Summit, Saddle Mountain State Park, Clatsop County, Oregon, August 23, 2014.

164   Moss's Elfin, *Callophrys mossii*. Summit, Saddle Mountain State Park, Clatsop County, Oregon, May 17, 2014.

165   Falcate Orangetip, *Anthocharis midea*. Douthat State Park, Bath County, Virginia, May 11, 2013.

166   Carolina satyr, *Hermeuptychia sosybius*. Vacant lot, Ebro, Washington County, Florida, April 16, 2014.

167   Palamedes Swallowtail, *Papilio palamedes*. Vacant lot, Ebro, Washington County, Florida, April 16, 2014.

168   Lindsey's Skipper, *Hesperia lindseyi*. Eastern portion, Cascade-Siskiyou National Monument, Jackson County, Oregon, May 29, 2014.

169   Mexican Yellow, *Eurema mexicana*. Chisos Basin, Big Bend National Park, Brewster County, Texas, April 26, 2014.

170   Vesta Crescent, *Phyciodes vesta*. Chisos Basin, Big Bend National Park, Brewster County, Texas, April 26, 2014.

171   Monarch, *Danaus plexippus*. Green Ridge near Camp Sherman, Jefferson County, Oregon, September 21, 2014.

172   Hydaspe Fritillary, *Speyeria hydaspe*. Road west from Big Summit Prairie, Ochoco National Forest, Crook County, Oregon, July 16, 2013.

173   Atlantis Fritillaries, *Speyeria atlantis*. Réserve Faunique La Vérendrye, Quebec, August 4, 2008.

174   Cabbage White, *Pieris rapae*. Twin Lakes Campground, Allegheny National Forest, Elk County, Pennsylvania, August 17, 2008.

174   Dun Skipper, *Euphyes vestris*. Twin Lakes Campground, Allegheny National Forest, Elk County, Pennsylvania, August 17, 2008.

174   Meadow Fritillary, *Boloria bellona*. Twin Lakes Campground, Allegheny National Forest, Elk County, Pennsylvania, August 17, 2008.

174   Red-spotted Purple form of White Admiral, *Limenitis arthemis astyanax*. Twin Lakes Campground, Allegheny National Forest, Elk County, Pennsylvania, August 17, 2008.

174   Summer Azure, *Celastrina neglecta*. Twin Lakes Campground, Allegheny National Forest, Elk County, Pennsylvania, August 17, 2008.

174   Aphrodite Fritillaries, *Speyeria aphrodite;* Eastern Tiger Swallowtail, *Papilio glaucus*. Twin Lakes Campground, Allegheny National Forest, Elk County, Pennsylvania, August 17, 2008.

175   Monarch, *Danaus plexippus*. Twin Lakes Campground, Allegheny National Forest, Elk County, Pennsylvania, August 17, 2008.

177   Two-tailed Tiger Swallowtail, *Papilio multicaudata;* Woodland Skippers, *Ochlodes*

sylvanoides. Cascade-Siskiyou National Monument, Jackson County, Oregon, August 9-10, 2003.

178   Mike Patterson, Oregon Silverspot, *Speyeria zerene hippolyta*. Mount Hebo, Siuslaw National Forest, Tillamook County, Oregon, August 19, 2014.

179   Oregon Silverspot, *Speyeria zerene hippolyta*. Mount Hebo, Siuslaw National Forest, Tillamook County, Oregon, August 12, 2002.

180   Painted Lady, *Vanessa cardui*. Myers home yard, Wahkiakum County, Washington, Summer 1998.

181   California Tortoiseshell, *Nymphalis californica*. Eagle Cap Wilderness, Wallowa County, Oregon, September 11, 2007.

182   Anicia Checkerspot, *Euphydryas anicia*. Swakane Canyon, Wenatchee National Forest, Wenatchee County, Washington, June 27±, 1999.

183   Sue Anderson, with Mourning Cloak, *Nymphalis antiopa*. Green Ridge near Camp Sherman, Jefferson County, Oregon, June 2, 2014.

184   Arctic Skipper, *Carterocephalus palaemon*. Metolius Preserve, Deschutes Land Trust Preserve near Camp Sherman, Jefferson County, Oregon, June 2, 2014.

185   Acmon/Lupine Blue, *Icaricia acmon/lupini*. Prairie Farm Meadow on Green Ridge near Camp Sherman, Jefferson County, Oregon, June 2, 2014.

186   Bramble Green Hairstreak, *Callophrys dumetorum;* Acmon Blue, *Icaricia acmon;* Echo Azure, *Celastrina echo;* Juniper Hairstreak, *Callophrys gryneus;* Brown Elfin, *Callophrys augustinus*. Green Ridge near Camp Sherman, Jefferson County, Oregon, June 2–3, 2014.

187   Great Arctic, *Oeneis nevadensis*. Tub Springs State Wayside, Jackson County, Oregon, May 30, 2014.

187   European Skipperling, *Thymelicus lineola*. Between Lake Nippigon and Hearst, Ontario, August 2, 2008.

188   Western Meadow Fritillary, *Boloria epithore*. Cow Meadow Camp, Cascade Lakes Highway, Deschutes National Forest, Deschutes County, Oregon, July 15, 2012.

189   Silver-bordered Fritillary, *Boloria selene*. Near east side of Big Summit Prairie, Ochoco National Forest, Crook County, Oregon, July 22, 2016.

191   Melissa Blues, *Plebejus melissa*. Crescent Mills, Plumas County, California, August 28, 1999.

192 Silvery Blue, *Glaucopsyche lygdamus*. Baldy Creek road, Cascade-Siskiyou National Monument, Jackson County, Oregon, May 30, 2014.

193 Silver-spotted Skipper, *Epargyreus clarus*. Eno Commons–Indigo Creek Trail, Durham County, North Carolina, October 6, 2016.

194 Large Marble, *Euchloe ausonides*. Hyatt Lake campground. Cascade-Siskiyou National Monument, Jackson County, Oregon, May 29, 2014.

195 Cabbage White, *Pieris rapae*. Myers home garden, Astoria, Clatsop County, Oregon, August 17, 2009.

196 Margined Whites, *Pieris marginalis*. Myers home garden, Wahkiakum County, Washington, Spring 2000.

197 David Branch, NABA count, Chumstick Mountain, Wenatchee National Forest, Chelan County, Washington, 1999.

197 Greenish Blue, *Icaricia saepiolus*. Vicinity of Allingham campgroud, Deschutes National Forest, Jefferson County, Oregon, June 27, 2013.

197 Sonora Skipper, *Polites sonora*. Grays River Weyerhaeuser Camp site, Wahkiakum County, Washington, June 1995.

197 Little Yellow on Fred Myers, *Pyrisitia lisa*. Huntley Meadows Park of Fairfax County, Virginia, September 1999.

197 Anicia Checkerspot, *Euphydryas anicia*. Swakane Canyon, Wenatchee County, Washington, June 26, 1999.

199 Pale Tiger Swallowtail, *Papilio eurymedon*, with Ann Potter. Flying L Ranch, Klickitat County, Washington, June 1995.

199 Cabbage White, *Pieris rapae*, with Andrew Emlen and Connor Emlen-Petterson. Alcyon Farm, Middle Valley, Skamokawa, Wahkiakum County, Washington, July 11, 2001.

200 Paul Hammond. Northwest Lepidopterists Society annual meeting, Oregon State Arthropod Collection, Corvallis, Oregon, October 28, 2006.

201 Pipevine Swallowtail, *Battus philenor*. Southeast of Del Rio near the Rio Grande, Maverick or Kinney County, Texas, April 24, 2014.

202–203 Hedgerow Hairstreak, *Satyrium saepium*. Santiam Pass, Deschutes National Forest, Jefferson County, Oregon, July 24, 2009.

204 Empress Leilia, *Asterocampa leilia*. Saguaro National Park, east unit, Pima County, Arizona, August 3–4, 2000.

205   Indra Swallowtail, *Papilio indra*. Mack's Canyon, Deschutes River National Recreation Area, Bureau of Land Management, Sherman County (Union County for the bluffs across the Deschutes), Oregon, April 24, 2013.

206   Gulf Fritillary, *Agraulis vanillae*. Spanish Harbor Wayside Park, Florida Department of Transportation, Monroe County, Florida, April 8, 2014.

207   Likely Zerene Fritillary, *Speyeria zerene*. Near Camp Sherman between U.S. route 20 and Green Ridge, Jefferson County, Oregon, September 20, 2014.

208   Hydaspe Fritillary, *Speyeria hydaspe*. Highway SR-4 right of way just west of KM summit, Wahkiakum County, Washington, August 4, 2013.

215   Eastern Tiger Swallowtail, *Papilio glaucus,* and Cindi Bailey. Reedy River Falls Park, City of Greenville, Greenville County, South Carolina, April 4, 2014.

216   Tyler Joki and Bob Pyle, butterflying class. Flying L Ranch, Klickitat County, Washington, June 1995.

217   Frederick Donahoe and Sachem Skipper, *Atalopedes campestris*. Home yard in Falls Church, Fairfax County, Virginia, August 22, 2010.

225   Boisduval's Blue, *Icaricia icarioides*. River Trail near mouth, Deschutes River State Recreation Area, Sherman County, Oregon, May 18, 2012.

227   Monarch, *Danaus plexippus*. Farm yard in Hop Bottom, Susquehanna County, Pennsylvania, August 13, 2008.

About the Author

LIFE AND PHOTOGRAPHY

My scientist father and artist mother raised me on the beaches, mountains, and deserts of southern California. I remember lying on Will Rogers State Park's beaches, studying sand grains an inch from my eye, each a different color and shape. During family outings to the Mojave, I felt the earth turn under the sky, and in the clear nights saw our position among the stars. I moved north to attend the University of California at Berkeley expecting to become a scientist, and left with degrees in mathematics and an abiding love for and commitment to fine art photography. What had happened? Nothing original—I had found a book by Ansel Adams, *These We Inherit: The Parklands of America*. Then Dave Bohn at the ASUC Studio supported my newfound interest with superb instruction.

In 1970 the romantic "back to the land" movement took me still farther north, to a clearing in the woods on Washington's rainforest coast, one ridge from the Columbia River. Settling in to a hundred inches of rain a year, I learned the seasons of garden and firewood. Trees, birds, butterflies, and all things wild became the stuff of daily life. In 2000 I crossed the Columbia to live in Astoria, Oregon—still under the same skies of the Columbia-Pacific region.

In the mid-1970s I began photographing natural history, starting with remnants of ancient forests. Landscapes, both classic sense-of-place and more abstract, remain of great interest.

Forty years of teaching college photography was a great satisfaction. Helping people find their own voices to explore the joys and challenges of life is a real highlight. Figuring out how to explain the technological and artistic issues increased my own understanding and helped my own work. Art and natural history photography and marriage make a good life. As this book was being composed, my wife and I moved to Ashland, Oregon, which is much better butterfly territory.

ENGAGING BUTTERFLIES

Like many children I chased butterflies. Mother made a net and Dad brought chloroform home from his research lab—we were all more casual with chemicals in the 1950s—so a few specimens got pinned. I was entranced by huge green caterpillars striped with black and yellow dots, found in fennel growing in a vacant lot. But mostly I went on the fool's errand of trying to hit skippers with a squirt gun, and succeeded not once. Then my interests drifted on to math, violin, surf, and girls.

Decades later Bob Pyle moved to Gray's River near my home in the woods, and in 1986 I resumed watching and netting butterflies. By 1995 I was ready for an additional photographic subject with new equipment, more mobile and spontaneous than my accustomed tripod-and-meditation camerawork, and this book is the result.

Index